Electronics Transistor Basics

FIRST EDITION

BY PRASUN BARUA

ABOUT

Welcome to Electronics Transistor Basics! This is a nonfiction science book which contains various topics on basics of transistor. A transistor is an electronic component that can be utilized in a variety of electronic devices because it can be used in circuits to either amplify or switch electrical signals or power. Two PN diodes connected back-to-back form a transistor. It has emitter, base, and collector terminals as its three terminals. The fundamental principle of a transistor is that it enables you to modify the intensity of a much smaller current that is flowing through a second channel in order to control the flow of current through one channel. Transistors are used extensively in daily life. They serve a purpose in almost every type of circuit in which they are utilized. Switches, amplifiers, oscillators, modulators, detectors, BJT, FET etc. are some of applications of transistors. This book contains various topics such as Bipolar Transistor, NPN Transistor, PNP Transistor, Transistor As A Switch, Junction Field Effect Transistor, MOSFET, MOSFET As A Switch, Darlington Transistors, Open Collector Outputs, FET Current Source and Transistor In A Brief. This is the first edition of the book. Thanks for reading the book.

TABLE OF CONTENTS

A semiconductor device called a bipolar junction transistor can be used for switching or amplification. Unlike semiconductor diodes, which have a single, straightforward pn-junction formed by two pieces of semiconductor material. An additional layer of semiconductor material is used in the bipolar transistor to create a device with amplifier-like qualities.

The result of connecting two separate signal diodes back-to-back in series will be two PN-junctions that share a common Positive (P) or Negative (N) terminal. These two diodes combine to make the Bipolar Junction Transistor, or BJT, which is a three-layer, two-junction, three-terminal device.

Transistors are three terminal active devices made of various semiconductor materials that, when applied with a modest signal voltage, can function as either an insulator or a conductor. The transistor may do either "switching" (digital electronics) or "amplification" due to its capacity to switch between these two states (analogue electronics). Consequently, bipolar transistors can function in one of three areas:

- Active Region – the transistor operates as an amplifier and $Ic = \beta*Ib$
- Saturation – the transistor is "Fully-ON" operating as a switch and $Ic = I(saturation)$
- Cut-off – the transistor is "Fully-OFF" operating as a switch and $Ic = 0$

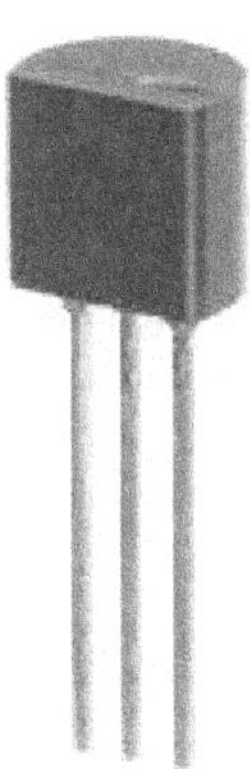

A typical Bipolar Transistor

The two terms Transfer Varistor, which characterize their manner of functioning in the early days of electronics development, are combined to form the word transistor. PNP and NPN, which roughly represent the physical arrangement of the P-type and N-type semiconductor materials from which they are manufactured, are the two fundamental forms of bipolar transistor fabrication.

The basic design of a bipolar transistor consists of two PN-junctions that result in three connected terminals; each terminal has a name to distinguish it from the other two. The Emitter (E), the Base (B), and the Collector (C) are the names and symbols for these three terminals, respectively.

Bipolar transistors serve as current-controlled switches by limiting the amount of current that flows through them from the Emitter to the Collector terminals in proportion to the biasing voltage that is provided to their base terminal.

The behavior of a transistor is controlled by a considerably greater collector current, which is created by a modest current flowing into the base terminal. The two transistor kinds, PNP and NPN, operate on the exact same concept; the only variations are in the biasing of the transistors and the polarity of the power supply for each type.

Construction of Bipolar Transistor

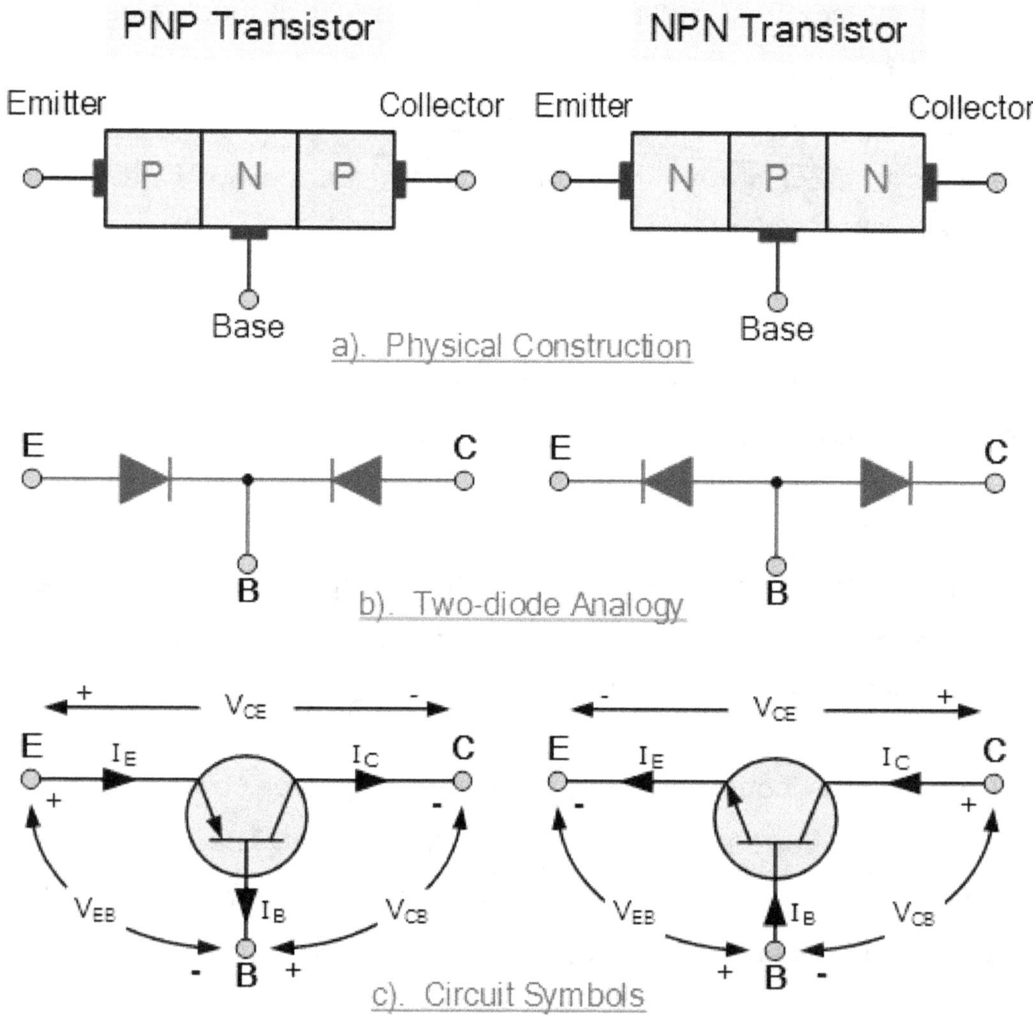

Both the PNP and NPN bipolar transistor's construction and circuit representations are provided above, with the arrow in the circuit symbol always indicating the direction of "conventional current flow" between the base terminal and emitter terminal. Similar to the conventional diode sign, the arrow always points from the positive P-type region to the negative N-type region for both types of transistors.

Configurations of Bipolar Transistors

As a three-terminal device, the bipolar transistor can be connected to an electronic circuit in one of three ways, with one terminal serving as the common terminal for the input and output signals. Since the static properties of the transistor differ depending on the circuit configuration, each way of connection reacts to its input signal differently inside a circuit.

- Common Base Configuration – has Voltage Gain but no Current Gain.
- Common Emitter Configuration – has both Current and Voltage Gain.
- Common Collector Configuration – has Current Gain but no Voltage Gain.

The configuration of the Common Base (CB)

In the Common Base or grounded base arrangement, as the name implies, the BASE connection is shared by the input signal AND the output signal. The base and emitter terminals of the transistor receive the input signal, and the base and collector terminals receive the matching output signal, as indicated. The base terminal can be linked to a fixed reference voltage point or is grounded.

In other words, the common base configuration "attenuates" the input signal. The input current flowing into the emitter is quite large because it is the sum of the base current and collector current, respectively. As a result, the collector current output is less than the emitter current input, resulting in a current gain for this type of circuit of "1" (unity) or less.

The Common Base Transistor Circuit

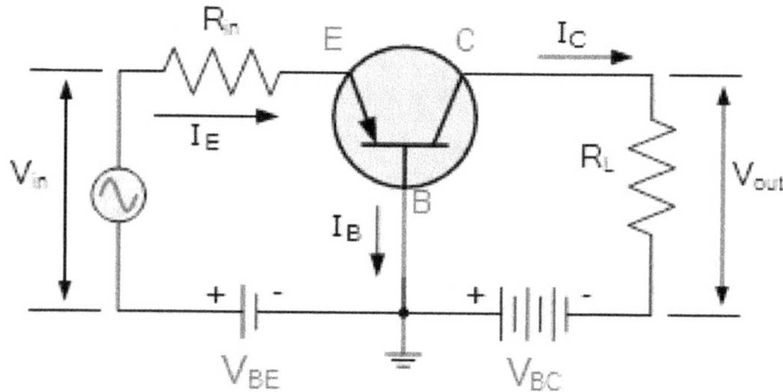

The signal voltages V_{in} and V_{out} are "in-phase" in this amplifier design, which is a non-inverting voltage amplifier circuit. Due to its exceptionally high voltage gain characteristics, this sort of transistor configuration is not very common.

Its output properties are those of an illuminated photo-diode, but its input characteristics are those of a forward biased diode. Additionally, this particular configuration of bipolar transistors has a high ratio of "load" resistance (R_L) to "input" resistance (R_{in}), which gives it a value of "Resistance Gain." Therefore, the following is the voltage gain (A_v) for a common base configuration:

Common Base Voltage Gain

$$A_V = \frac{Vout}{Vin} = \frac{I_C \times R_L}{I_E \times R_{IN}}$$

Here, Ic/Ie is the current gain, alpha (α) and RL/Rin is the resistance gain. The common base circuit is typically used in single stage amplifier circuits such as microphone pre-amplifier or radio frequency (Rf) amplifiers due to its very good high frequency response.

Configuration of the Common Emitter (CE)

The input signal is applied between the base and the emitter in the common emitter or grounded emitter configuration, and the output is obtained, as indicated, from between the collector and the emitter. This arrangement, which exemplifies the "standard" way to connect bipolar transistors, is the most popular circuit for transistor-based amplifiers.

Of the three bipolar transistor topologies, the common emitter amplifier produces the highest current and power gain. This is mostly due to the fact that the output impedance is HIGH because it is drawn from a reverse biased PN-junction while the input impedance is LOW because it is connected to a forward biased PN-junction.

The Common Emitter Amplifier Circuit

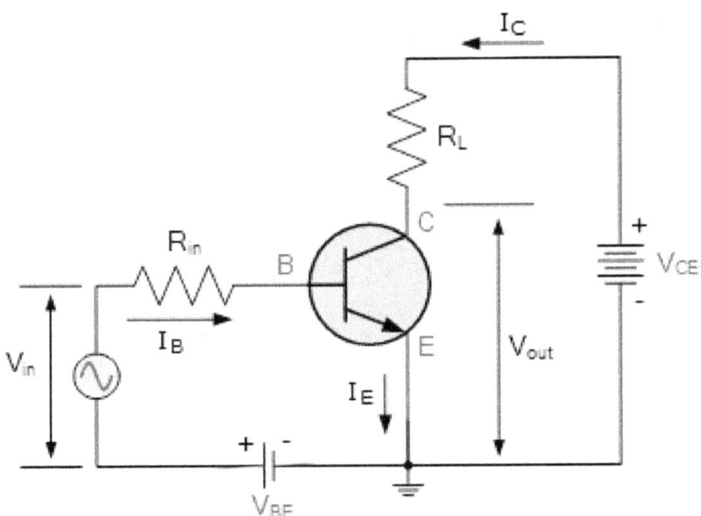

In this type of configuration, the current flowing out of the transistor must be equal to the currents flowing into the transistor as the emitter current is given as Ie = Ic + Ib.

Since the load resistance (R_L) is connected in series with the collector, the current gain of the common emitter transistor configuration is quite large as it is the ratio of Ic/Ib. A transistors current gain is given the Greek symbol of Beta, (β).

Since the emitter current for a common emitter configuration is defined as Ie = Ic + Ib, the ratio of Ic/Ie is called Alpha, given the Greek symbol of α. Note: that the value of Alpha will always be less than unity.

As the electrical relationship between these three currents, Ib, Ic and Ie is determined by the physical construction of the transistor itself, any small change in the base current (Ib), will result in a much larger change in the collector current (Ic).

The current in the emitter-collector circuit will thereafter be controlled by little variations in the base's current. For the majority of general-purpose transistors, Beta typically ranges from 20 to 200. Accordingly, one electron will flow from the base terminal for every 100 electrons that move between the emitter-collector terminal and base terminal of a transistor with a beta value of, say, 100.

By combining the expressions for both Alpha, α and Beta, β the mathematical relationship between these parameters and therefore the current gain of the transistor can be given as:

$$\text{Alpha,}(\alpha) = \frac{I_C}{I_E} \quad \text{and} \quad \text{Beta,}(\beta) = \frac{I_C}{I_B}$$

$$\therefore I_C = \alpha.I_E = \beta.I_B$$

$$\text{as:} \quad \alpha = \frac{\beta}{\beta + 1} \qquad \beta = \frac{\alpha}{1 - \alpha}$$

$$I_E = I_C + I_B$$

Here, "Ic" is the current flowing into the collector terminal, "Ib" is the current flowing into the base terminal and "Ie" is the current flowing out of the emitter terminal.

Compared to the conventional base design, this sort of bipolar transistor configuration has a higher input impedance, current gain, and power gain, but a far lower voltage gain. An inverting amplifier circuit has a common emitter arrangement. This indicates a 180° phase-shift between the output signal and the input voltage signal in the output signal.

Configuration of the Common Collector (CC)

The collector terminal is shared by the input and output in the common collector or grounded collector configuration because the collector is connected to ground via the supply. The output signal is obtained as illustrated from across the emitter load resistor, while the input signal is linked directly to the base terminal.

Common names for this kind of setup include voltage follower and emitter follower circuits. The common collector, or emitter follower arrangement has a very high input impedance in the hundreds of thousands of ohms range and a comparatively low output impedance, making it ideal for impedance matching applications.

The Common Collector Transistor Circuit

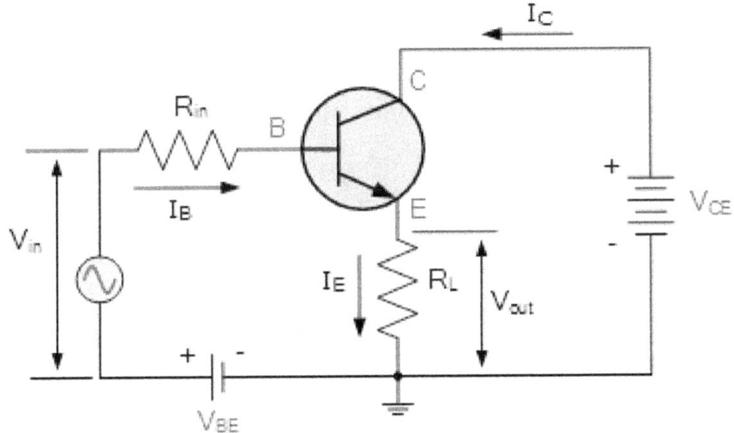

The common emitter arrangement offers a current gain that is about equal to the transistor's β value. In the usual collector setup, however, the load resistance is linked in series with the emitter terminal, thus its current is equal to the emitter current. Since the emitter current is the sum of the collector and base currents, the load resistance in this sort of transistor design also has the collector and base input currents flowing through it. The current gain of the circuit is therefore given as:

The Common Collector Current Gain

$$I_E = I_C + I_B$$

$$A_i = \frac{I_E}{I_B} = \frac{I_C + I_B}{I_B}$$

$$A_i = \frac{I_C}{I_B} + 1$$

$$A_i = \beta + 1$$

This is a non-inverting bipolar transistor design since the signal voltages V_{in} and V_{out} are "in-phase." The voltage gain of the conventional collector design is approximately "1." (unity gain). Because the voltage gain is one, it might be called a voltage buffer.

The common collector transistor's load resistance receives both the base and collector currents, resulting in a substantial current gain (as with the common emitter arrangement) and good current amplification with very little voltage gain.

After reviewing the three different types of bipolar transistor topologies, the following table summarizes the numerous relationships between the transistor's separate DC currents flowing through each leg and its DC current gains.

Relationship between DC Currents and Gains

$$I_E = I_B + I_C \qquad \alpha = \frac{I_C}{I_E} = \frac{\beta}{1+\beta}$$

$$I_C = I_E - I_B$$

$$I_B = I_E - I_C \qquad \beta = \frac{I_C}{I_B} = \frac{\alpha}{1-\alpha}$$

$$I_B = \frac{I_C}{\beta} = \frac{I_E}{1+\beta} = I_E(1-\alpha)$$

$$I_C = \beta.I_B = \alpha.I_E \qquad I_E = \frac{I_C}{\alpha} = I_B(1+\beta)$$

Although we have looked at NPN Bipolar Transistor setups here, PNP transistors are just as valid to utilize in each configuration because the calculations will all be the same, as for the amplified signal not being inverted. The only change will be in the polarity of the voltage and current.

Bipolar Transistor In A Brief

In conclusion, the behavior of the bipolar transistor in each of the aforementioned circuit configurations is very different, and this results in different circuit characteristics in terms of input impedance, output impedance, and gain, whether this is voltage gain, current gain, or power gain. This is summarized in the table below.

Configurations of Bipolar Transistor

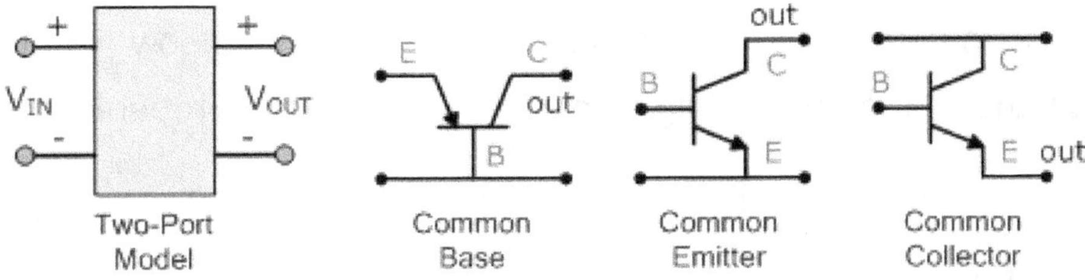

the following table's generalized features of the various transistor configurations:

Characteristic	Common Base	Common Emitter	Common Collector
Input Impedance	Low	Medium	High
Output Impedance	Very High	High	Low
Phase Shift	$0°$	$180°$	$0°$
Voltage Gain	High	Medium	Low
Current Gain	Low	Medium	High
Power Gain	Low	Very High	Medium

CHAPTER-2: NPN TRANSISTOR

NPN Transistors are three-terminal, three-layer components that can be used as electronic switches or amplifiers. The ordinary bipolar transistor, or BJT, is available in two fundamental configurations. An NPN configuration (Negative-Positive-Negative) and a PNP configuration (Positive-Negative-Positive) are both possible. That is, there are two different types of transistors: NPN and PNP.

The NPN transistor configuration is the one that is most frequently utilized. Additionally, we discovered that the bipolar transistor's junctions can be biased in one of three ways: common base, common emitter, or common collector. Here, we will take a closer look at the "Common Emitter" arrangement using a bipolar NPN transistor. Below is an illustration of how an NPN transistor is built, along with information on the properties of the transistors' current flow.

Configuration of A Bipolar NPN Transistor

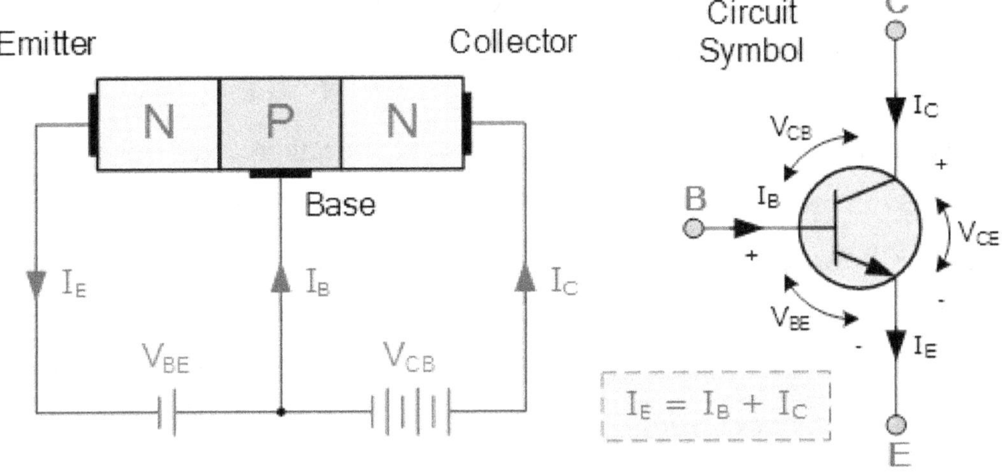

(Note: Arrow defines the emitter and conventional current flow, "out" for a Bipolar NPN Transistor.)

The construction and terminal voltages for a bipolar NPN transistor are shown above. The voltage between the Base and Emitter (V_{BE}), is positive at the Base and negative at the Emitter because for an NPN transistor, the Base terminal is always positive with respect to the Emitter. Also the Collector supply voltage is positive with respect to the Emitter (V_{CE}). So for a bipolar NPN transistor to conduct the Collector is always more positive with respect to both the Base and the Emitter.

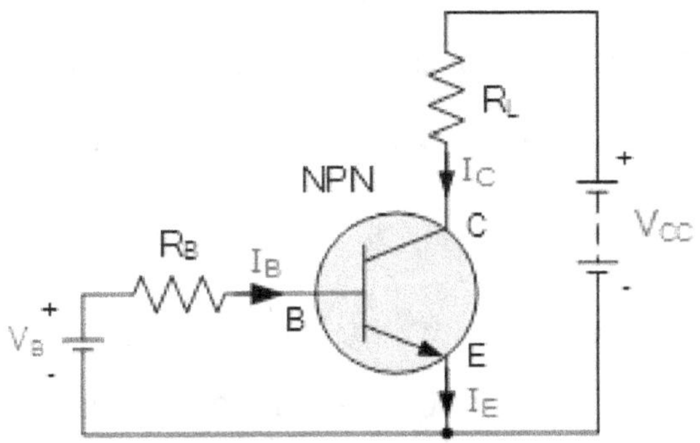

NPN Transistor Connection

Voltage sources are connected to an NPN transistor as shown. The Collector is connected to the supply voltage V_{CC} via the load resistor, RL which also acts to limit the maximum current flowing through the device. The Base supply voltage V_B is connected to the Base resistor R_B, which again is used to limit the maximum Base current.

Therefore, in a NPN Transistor it is the movement of negative current carriers (electrons) through the Base region that constitutes transistor action, since these mobile electrons provide the link between the Collector and Emitter circuits. This link between the input and output circuits is the main feature of transistor action because the transistors amplifying properties come from the consequent control which the Base exerts upon the Collector to Emitter current.

Transistor is a current operated device (Beta model) and that a large current (Ic) flows freely through the device between the collector and the emitter terminals when the transistor is switched "fully-ON". However, this only happens when a small biasing current (Ib) is flowing into the base terminal of the transistor at the same time thus allowing the Base to act as a sort of current control input.

The current in a bipolar NPN transistor is the ratio of these two currents (Ic/Ib), called the *DC Current Gain* of the device and is given the symbol of hfe or nowadays Beta, (β).

The value of β can be large up to 200 for standard transistors, and it is this large ratio between Ic and Ib that makes the bipolar NPN transistor a useful amplifying device when used in its active region as Ib provides the input and Ic provides the output. Note that Beta has no units as it is a ratio.

Also, the current gain of the transistor from the Collector terminal to the Emitter terminal, Ic/Ie, is called Alpha, (α), and is a function of the transistor itself (electrons diffusing across the junction). As the emitter current Ie is the sum of a very small base current plus a very large collector current, the value of alpha (α), is very close to unity, and for a typical low-power signal transistor this value ranges from about 0.950 to 0.999

α and β Relationship in a NPN Transistor

$$\text{DC Current Gain} = \frac{\text{Output Current}}{\text{Input Current}} = \frac{I_C}{I_B}$$

$$I_E = I_B + I_C \;\dots\; (KCL) \quad \text{and} \quad \frac{I_C}{I_E} = \alpha$$

$$\text{Thus}: \; I_B = I_E - I_C$$

$$I_B = I_E - \alpha I_E$$

$$I_B = I_E\left(1 - \alpha\right)$$

$$\therefore \; \beta = \frac{I_C}{I_B} = \frac{I_C}{I_E\left(1 - \alpha\right)} = \frac{\alpha}{1 - \alpha}$$

By combining the two parameters α and β we can produce two mathematical expressions that gives the relationship between the different currents flowing in the transistor.

$$\alpha = \frac{\beta}{\beta + 1} \quad \text{or} \quad \alpha = \beta(1 - \alpha)$$

$$\beta = \frac{\alpha}{1 - \alpha} \quad \text{or} \quad \beta = \alpha(1 + \beta)$$

$$\text{If} \quad \alpha = 0.99 \qquad \beta = \frac{0.99}{0.01} = 99$$

The values of Beta vary from about 20 for high current power transistors to well over 1000 for high frequency low power type bipolar transistors. The value of Beta for most standard NPN transistors can be found in the manufactures data sheets but generally range between 50 – 200.

The equation above for Beta can also be re-arranged to make Ic as the subject, and with a zero base current (Ib = 0) the resultant collector current Ic will also be zero, (β*0). Also when the base current is high the corresponding collector current will also be high resulting in the base current controlling the collector current. One of the most important properties of the Bipolar Junction Transistor is that a small base current can control a much larger collector current. Consider the following example.

Example-1 of NPN Transistor

A bipolar NPN transistor has a DC current gain, (Beta) value of 200. Calculate the base current Ib required to switch a resistive load of 4mA.

$$I_B = \frac{I_C}{\beta} = \frac{4 \times 10^{-3}}{200} = 20\mu A$$

So, β = 200, Ic = 4mA and Ib = 20µA.

The collector voltage, (Vc) must be greater and positive with respect to the emitter voltage, (Ve) to allow current to flow through the transistor between the collector-emitter junctions. Also, there is a voltage drop between the Base and the Emitter terminal of about 0.7V (one diode volt drop) for silicon devices as the input characteristics of an NPN Transistor are of a forward biased diode. The base voltage, (Vbe) of a NPN transistor must be greater than this 0.7V otherwise the transistor will not conduct with the base current given as.

$$I_b = V_b - V_{be} \: / \: R_b$$

Here, Ib is the base current, Vb is the base bias voltage, Vbe is the base-emitter volt drop (0.7v) and Rb is the base input resistor. Increasing Ib, Vbe slowly increases to 0.7V but Ic rises exponentially.

Example-2 of NPN Transistor

An NPN Transistor has a DC base bias voltage, Vb of 10v and an input base resistor, Rb of 100kΩ. Calculate the value of the base current into the transistor.

$$I_B = \frac{V_B - V_{BE}}{R_B} = \frac{10 - 0.7}{100k\Omega} = 93\mu A$$

So, Ib = 93µA.

Configuration of the Common Emitter

Bipolar NPN Transistors can be used in their active region to create a circuit that will amplify any small AC signal applied to its Base terminal with the Emitter grounded, in addition to being used as a semiconductor switch to turn load currents "ON" or "OFF" by controlling the Base signal to the transistor in either its saturation or cut-off regions.

A single stage common emitter amplifier, an inverting amplifier circuit, is created if an appropriate DC "biasing" voltage is initially given to the transistor's base terminal, allowing it to always function within its linear active area.

A Class A amplifier is one such Common Emitter Amplifier arrangement of an NPN transistor. When the transistor's Base terminal is biased in order to forward bias the Base-emitter junction, the circuit is said to be operating as a "Class A amplifier."

The transistor is therefore always midway between its cut-off and saturation zones, enabling the transistor amplifier to faithfully recreate both the positive and negative sides of any AC input signal overlaid onto this DC biasing voltage.

Only one half of the input waveform would be amplified absent this "Bias Voltage." There are various uses for this simple emitter amplifier design using an NPN transistor, but it is frequently used in audio circuits like pre-amplifier and power amplifier stages.

With reference to the common emitter configuration shown below, a family of curves known as the Output Characteristics Curves, relates the output collector current, (I_c) to the collector voltage, (V_{ce}) when different values of Base current, (I_b). Output characteristics curves are applied to the transistor for transistors with the same β value.

A DC "Load Line" can also be drawn onto the output characteristics curves to show all the possible operating points when different values of base current are applied. It is necessary to set the initial value of Vce correctly to allow the output voltage to vary both up and down when amplifying AC input signals and this is called setting the operating point or Quiescent Point, Q-point for short and this is shown below.

Single Stage Common Emitter Amplifier Circuit

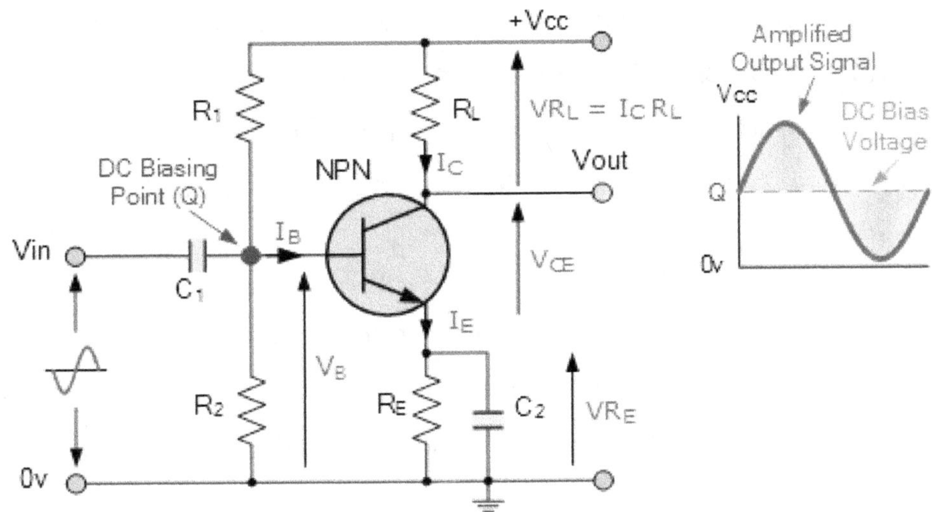

Output Characteristics Curves of a Typical Bipolar Transistor

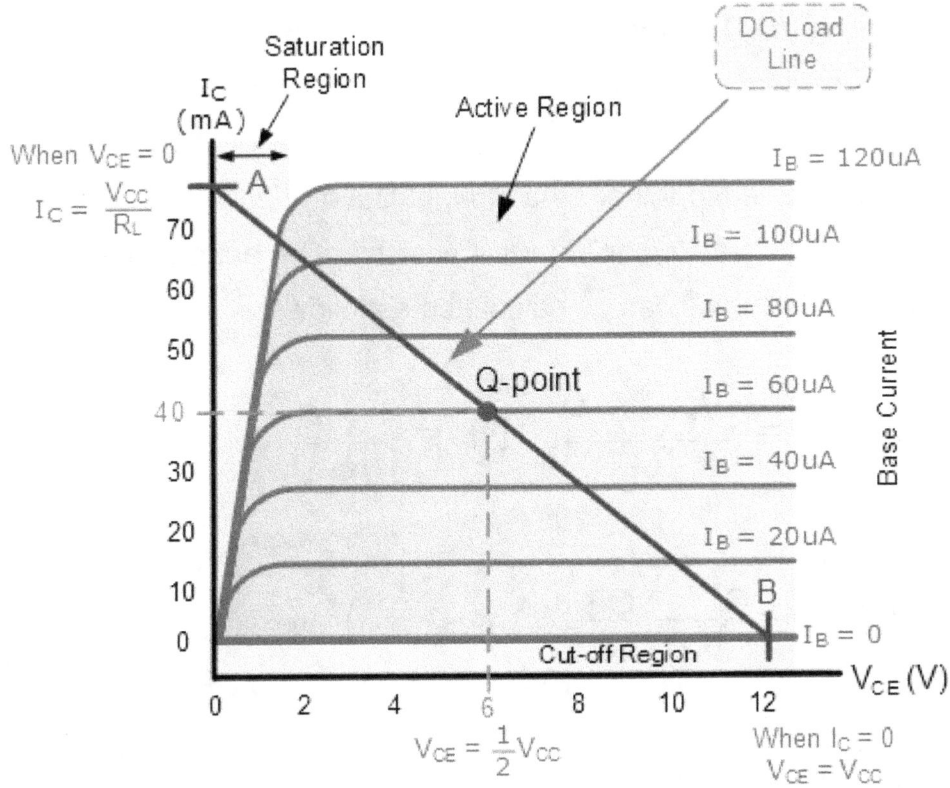

The most important factor to notice is the effect of Vce upon the collector current Ic when Vce is greater than about 1.0 volts. We can see that Ic is largely unaffected by changes in Vce above this value and instead it is almost entirely controlled by the base current, Ib. When this happens we can say then that the output circuit represents that of a "Constant Current Source".

It can also be seen from the common emitter circuit above that the emitter current Ie is the sum of the collector current, Ic and the base current, Ib, added together so we can also say that Ie = Ic + Ib for the common emitter (CE) configuration.

By using the output characteristics curves in our example above and also Ohm´s Law, the current flowing through the load resistor, (R_L), is equal to the collector current, Ic entering the transistor which in turn corresponds to the supply voltage, (Vcc) minus the voltage drop between the collector and the emitter terminals, (Vce) and is given as:

$$\text{Collector Current, } I_C = \frac{V_{CC} - V_{CE}}{R_L}$$

Also, a straight line representing the **Dynamic Load Line** of the transistor can be drawn directly onto the graph of curves above from the point of "Saturation" (A) when Vce = 0 to the point of "Cut-off" (B) when Ic = 0 thus giving us the "Operating" or **Q-point** of the transistor. These two points are joined together by a straight line and any position along this straight line represents the "Active Region" of the transistor. The actual position of the load line on the characteristics curves can be calculated as follows:

$$\text{When: } \left(V_{CE} = 0 \right) \quad I_C = \frac{V_{CC} - 0}{R_L}, \quad I_C = \frac{V_{CC}}{R_L}$$

$$\text{When: } \left(I_C = 0 \right) \quad 0 = \frac{V_{CC} - V_{CE}}{R_L}, \quad V_{CC} = V_{CE}$$

The collector or output characteristics curves for **Common Emitter NPN Transistors** can be used to predict the Collector current, Ic, when given Vce and the Base current, Ib. A Load Line can also be constructed onto the curves to determine a suitable Operating or **Q-point** which can be set by adjustment of the base current. The slope of this load line is equal to the reciprocal of the load resistance which is given as: $-1/R_L$

We can define a **NPN Transistor** as being normally "OFF" but a small input current and a small positive voltage at its Base (B) relative to its Emitter (E) will turn it "ON" allowing a much large Collector-Emitter current to flow. NPN transistors conduct when Vc is much greater than Ve.

CHAPTER-3: PNP TRANSISTOR

The NPN Transistor device is the exact opposite of the PNP Transistor. In essence, compared to the prior NPN transistor, the two coupled diodes are flipped in this form of PNP transistor construction. This results in a Positive-Negative-Positive configuration, with the arrow defining the Emitter terminal in the PNP transistor sign pointing inwards.

Additionally, a PNP transistor has all of its polarities reversed, so that it "sources" current through its Base as opposed to an NPN transistor, which "sinks" current into its Base. The primary distinction between the two types of transistors is that PNP transistors' more significant carriers are holes, whereas NPN transistors' significant carriers are electrons.

In order to regulate a considerably bigger emitter-collector current, PNP transistors use a lower base current and a negative base voltage. To put it another way, the Emitter of a PNP transistor is more positive both in relation to the Base and the Collector. As seen below, two P-type semiconductor materials are placed either side of an N-type material to form a "PNP transistor."

Configuration of A PNP Transistor

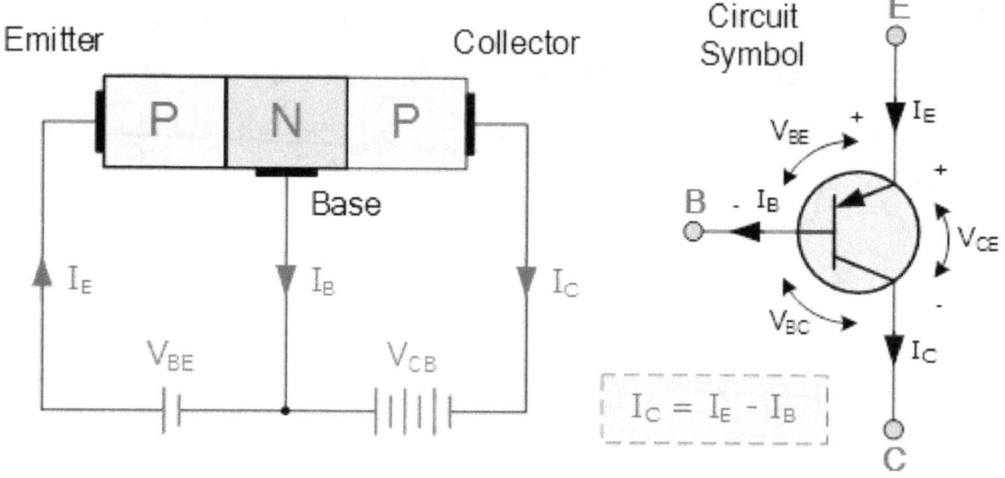

(Note: Arrow defines the emitter and conventional current flow, "in" for a PNP transistor.)

An NPN transistor's design and terminal voltages are depicted above. The PNP Transistor shares many traits with its NPN bipolar relatives, with the exception that for any of the three possible configurations—Common Base, Common Emitter, and Common Collector—the polarities (or biasing) of the current and voltage directions are reversed.

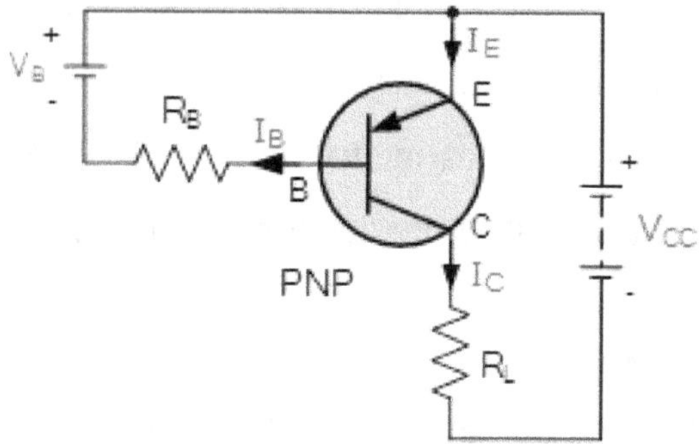

PNP Transistor Connection

The voltage between the Base and Emitter (V_{BE}), is now negative at the Base and positive at the Emitter because for a PNP transistor, the Base terminal is always biased negative with respect to the Emitter.

Also the Emitter supply voltage is positive with respect to the Collector (V_{CE}). So for a PNP transistor to conduct the Emitter is always more positive with respect to both the Base and the Collector.

The voltage sources are connected to a PNP transistor are as shown. This time the Emitter is connected to the supply voltage V_{CC} with the load resistor, RL which limits the maximum current flowing through the device connected to the Collector terminal. The Base voltage V_B which is biased negative with respect to the Emitter and is connected to the Base resistor R_B, which again is used to limit the maximum Base current.

To cause the Base current to flow in a PNP transistor the Base needs to be more negative than the Emitter (current must leave the base) by approx 0.7 volts for a silicon device or 0.3 volts for a germanium device with the formulas used to calculate the Base resistor, Base current or Collector current are the same as those used for an equivalent NPN transistor and is given as.

$$I_C = I_E - I_B$$

$$I_C = \beta.I_B \qquad I_B = \frac{I_C}{\beta}$$

It has been observed that the fundamental differences between a NPN Transistor and a PNP Transistor is the proper biasing of the transistors junctions as the current directions and voltage polarities are always opposite to each other. Therefore, for the circuit above: Ic = Ie − Ib as current must leave the Base.

Usually, the PNP transistor can replace NPN transistors in most electronic circuits, the only difference is the polarities of the voltages, and the directions of the current flow. PNP transistors can also be used as switching devices and an example of a PNP transistor switch is shown below.

A PNP Transistor Circuit

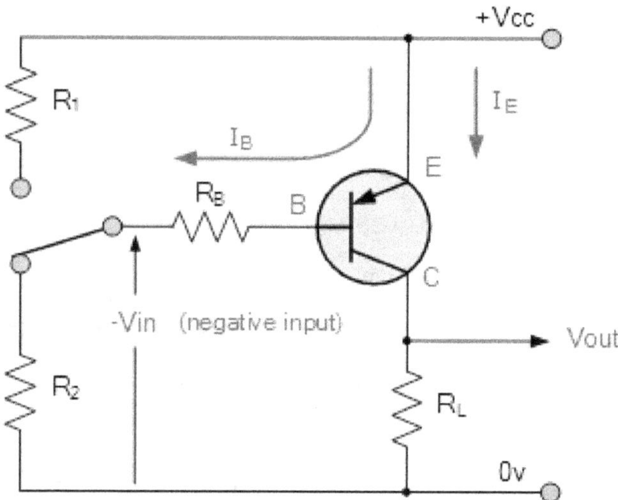

The Output Characteristics Curves for a PNP transistor look very similar to those for an equivalent NPN transistor except that they are rotated by 180° to take account of the reverse polarity voltages and currents, (that is for a PNP transistor, electron current flows out of the base and collector towards the battery). The same dynamic load line can be drawn onto the I-V curves to find the PNP transistors operating points.

Transistor Matching

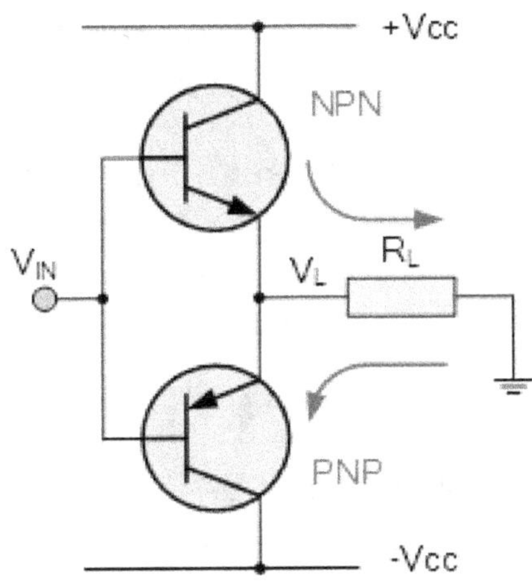

Complementary Transistors

What's the use of having a PNP transistor when there are so many NPN transistors that can be utilized as solid-state switches or amplifiers, you might wonder. It turns out that having two distinct transistor types, PNP and NPN, can be quite helpful for developing power amplifier circuits, like the Class B Amplifier.

In its output stage or in reversible H-Bridge motor control circuits, Class-B amplifiers use "Complementary" or "Matched Pair" (that is, one PNP and one NPN connected together) transistors. These transistors allow us to regulate the flow of current through the motor evenly in both directions at different times for forward and reverse motion.

A pair of corresponding NPN and PNP transistors with near identical characteristics to each other are called Complementary Transistors for example, a TIP3055 (NPN transistor) and the TIP2955 (PNP transistor) are good examples of complementary or matched pair silicon power transistors. They both have a DC current gain, Beta, (Ic/Ib) matched to within 10% and high Collector current of about 15A making them ideal for general motor control or robotic applications.

Also, class B amplifiers use complementary NPN and PNP in their power output stage design. The NPN transistor conducts for only the positive half of the signal while the PNP transistor conducts for negative half of the signal.

As a result, the output current is likely to be in the range of several amps, evenly distributed between the two complimentary transistors, allowing the amplifier to drive the necessary power through the load loudspeaker in both directions at the stated nominal impedance and power.

Identifying the PNP Transistor

We know that transistors are basically made up of two Diodes connected together back-to-back. By measuring the resistance between the emitter, base, and collector leads, we may use this analogy to determine whether a transistor is of the PNP type or NPN type. Six tests total with the anticipated resistance values in Ohms are obtained by using a multimeter to test each pair of transistor leads in both directions.

> - **Emitter-Base Terminals:** The Emitter to Base should act like a normal diode and conduct one way only.
> - **Collector-Base Terminals:** The Collector-Base junction should act like a normal diode and conduct one way only.
> - **Emitter-Collector Terminals:** The Emitter-Collector should not conduct in either direction.

Terminal Resistance Values for PNP and NPN Transistors

Between Transistor Terminals		PNP	NPN
Collector	Emitter	R_{HIGH}	R_{HIGH}
Collector	Base	R_{LOW}	R_{HIGH}
Emitter	Collector	R_{HIGH}	R_{HIGH}
Emitter	Base	R_{LOW}	R_{HIGH}
Base	Collector	R_{HIGH}	R_{LOW}
Base	Emitter	R_{HIGH}	R_{LOW}

We can say that a PNP transistor is typically "OFF," but that it can be made "ON" by applying a little output current and a negative voltage to the base (B) relative to the emitter (E) to cause a much larger Emitter-Collector current to flow. When Ve is significantly more than Vc, PNP transistors begin to operate.

To put it another way, a bipolar PNP transistor will only operate if the Base and Collector terminals are both negative in relation to the Emitter. Bipolar transistor switches are used in many different applications to turn a DC current "ON" or "OFF," whether it be for motors or relays that may need bigger currents at higher voltages, or for LEDs that just need a few milliamps of switching current at low DC voltages.

Transistor switches can be used to switch a low voltage DC device (e.g. LED's) ON or OFF by using a transistor in its saturated or cut-off state When used as an AC signal amplifier, the transistors Base biasing voltage is applied in such a way that it always operates within its "active" region, that is the linear part of the output characteristics curves are used. However, both the NPN & PNP type bipolar transistors can be made to operate as "ON/OFF" type solid state switch by biasing the transistors Base terminal differently operating the transistor as a switch.

Solid state switches are one of the main applications for the use of transistor to switch a DC output "ON" or "OFF". Some output devices, such as LED's only require a few milliamps at logic level DC voltages and can therefore be driven directly by the output of a logic gate. However, high power devices such as motors, solenoids or lamps, often require more power than that supplied by an ordinary logic gate so transistor switches are used.

If the circuit uses the **Bipolar Transistor as a Switch**, then the biasing of the transistor, either NPN or PNP is arranged to operate the transistor at both sides of the " I-V " characteristics curves we have seen previously.

The areas of operation for a transistor switch are known as the **Saturation Region** and the **Cut-off Region**. This means then that we can ignore the operating Q-point biasing and voltage divider circuitry required for amplification, and use the transistor as a switch by driving it back and forth between its "fully-OFF" (cut-off) and "fully-ON" (saturation) regions as shown below.

Operating Regions

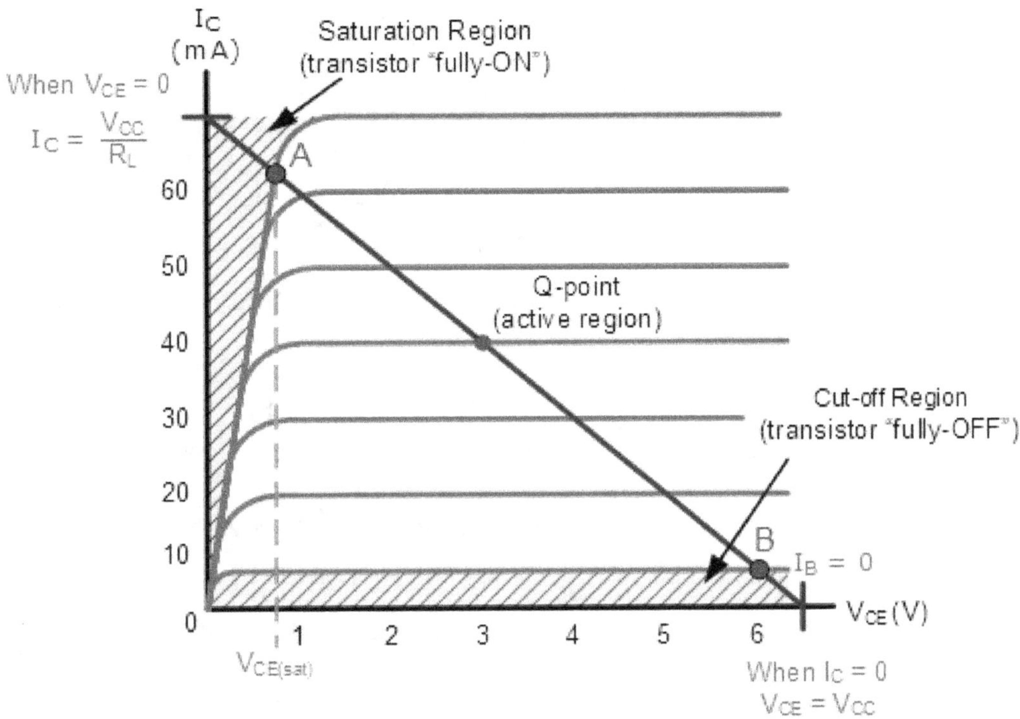

The pink shaded area at the bottom of the curves represents the "Cut-off" region while the blue area to the left represents the "Saturation" region of the transistor. Both these transistor regions are defined as:

1. Cut-off Region

Here the operating conditions of the transistor are zero input base current (I_B), zero output collector current (I_C) and maximum collector voltage (V_{CE}) which results in a large depletion layer and no current flowing through the device. Therefore the transistor is switched "Fully-OFF".

Cut-off Characteristics

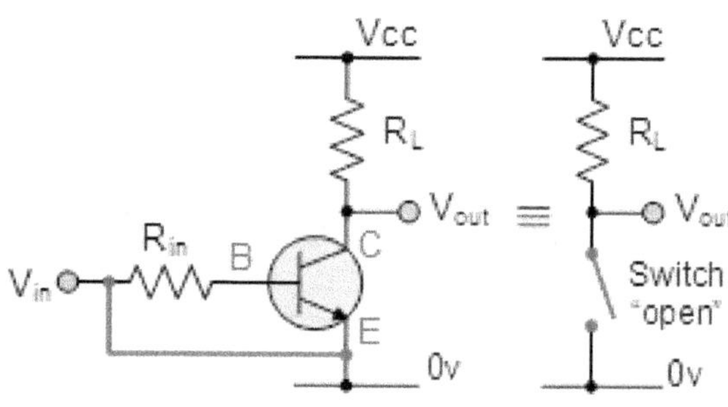

- The input and Base are grounded (0v)
- Base-Emitter voltage V_{BE} < 0.7v
- Base-Emitter junction is reverse biased
- Base-Collector junction is reverse biased
- Transistor is "fully-OFF" (Cut-off region)
- No Collector current flows (I_C = 0)
- $V_{OUT} = V_{CE} = V_{CC}$ = "1"
- Transistor operates as an "open switch"

Then we can define the "cut-off region" or "OFF mode" when using a bipolar transistor as a switch as being, both junctions reverse biased, V_B < 0.7v and I_C = 0. For a PNP transistor, the Emitter potential must be negative with respect to the Base.

2. Saturation Region

Here the transistor will be biased so that the maximum amount of base current is applied, resulting in maximum collector current resulting in the minimum collector emitter voltage drop which results in the depletion layer being as small as possible and maximum current flowing through the transistor. Therefore, the transistor is switched "Fully-ON".

Saturation Characteristics

- The input and Base are connected to V_{CC}
- Base-Emitter voltage $V_{BE} > 0.7v$
- Base-Emitter junction is forward biased
- Base-Collector junction is forward biased
- Transistor is "fully-ON" (saturation region)
- Max Collector current flows ($I_C = Vcc/R_L$)
- $V_{CE} = 0$ (ideal saturation)
- $V_{OUT} = V_{CE} =$ "0"
- Transistor operates as a "closed switch"

The "saturation region" or "ON mode" can be defined when using a bipolar transistor as a switch as being, both junctions forward biased, $V_B > 0.7v$ and I_C = Maximum. For a PNP transistor, the Emitter potential must be positive with respect to the Base.

The transistor operates as a "single-pole single-throw" (SPST) solid state switch. With a zero signal applied to the Base of the transistor it turns "OFF" acting like an open switch and zero collector current flows. With a positive signal applied to the Base of the transistor it turns "ON" acting like a closed switch and maximum circuit current flows through the device.

The simplest way to switch moderate to high amounts of power is to use the transistor with an open-collector output and the transistors Emitter terminal connected directly to ground. When used in this way, the transistors open collector output can thus "sink" an externally supplied voltage to ground thereby controlling any connected load.

An example of an NPN Transistor as a switch being used to operate a relay is given below. With inductive loads such as relays or solenoids a flywheel diode is placed across the load to dissipate the back EMF generated by the inductive load when the transistor switches "OFF" and so protect the transistor from damage. If the load is of a very high current or voltage nature, such as motors, heaters etc, then the load current can be controlled via a suitable relay as shown.

Basic NPN Transistor Switching Circuit

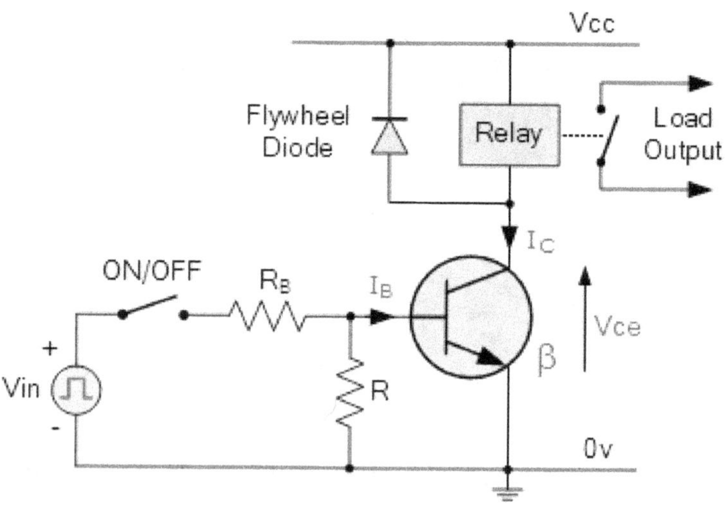

The circuit is similar to that of the *Common Emitter* circuit. The difference this time is that to operate the transistor as a switch the transistor needs to be turned either fully "OFF" (cut-off) or fully "ON" (saturated).

An ideal transistor switch would have infinite circuit resistance between the Collector and Emitter when turned "fully-OFF" resulting in zero current flowing through it and zero resistance between the Collector and Emitter when turned "fully-ON", resulting in maximum current flow.

In practice when the transistor is turned "OFF", small leakage currents flow through the transistor and when fully "ON" the device has a low resistance value causing a small saturation voltage (V_{CE}) across it. Even though the transistor is not a perfect switch, in both the cut-off and saturation regions the power dissipated by the transistor is at its minimum.

In order for the Base current to flow, the Base input terminal must be made more positive than the Emitter by increasing it above the 0.7 volts needed for a silicon device. By varying this Base-Emitter voltage V_{BE}, the Base current is also altered and which in turn controls the amount of Collector current flowing through the transistor as previously discussed.

When maximum Collector current flows the transistor is said to be Saturated. The value of the Base resistor determines how much input voltage is required and corresponding Base current to switch the transistor fully "ON".

Transistor as a Switch Example-1

Using the transistor values of: β = 200, Ic = 4mA and Ib = 20uA, determine the value of the Base resistor (Rb) required to switch the load fully "ON" when the input terminal voltage exceeds 2.5v.

$$R_B = \frac{V_{in} - V_{BE}}{I_B} = \frac{2.5v - 0.7v}{20 \times 10^{-6}} = 90k\Omega$$

The next lowest preferred value is: 82kΩ, this guarantees the transistor switch is always saturated.

Transistor as a Switch Example-2

Again, using the same values, find the minimum Base current required to turn the transistor "fully-ON" (saturated) for a load that requires 200mA of current when the input voltage is increased to 5.0V. Also calculate the new value of Rb.

Transistor Base current:

$$I_B = \frac{I_C}{\beta} = \frac{200mA}{200} = 1mA$$

Transistor Base resistance:

$$R_B = \frac{V_{in} - V_{BE}}{I_B} = \frac{5.0v - 0.7v}{1 \times 10^{-3}} = 4.3k\Omega$$

Transistor switches are used for a wide variety of applications such as interfacing large current or high voltage devices like motors, relays or lamps to low voltage digital IC's or logic gates like AND gates or OR gates.

Here, the output from a digital logic gate is only +5v but the device to be controlled may require a 12 or even 24 volts supply. Or the load such as a DC Motor may need to have its speed controlled using a series of pulses (Pulse Width Modulation). transistor switches will allow us to do this faster and more easily than with conventional mechanical switches.

Digital Logic Transistor Switch

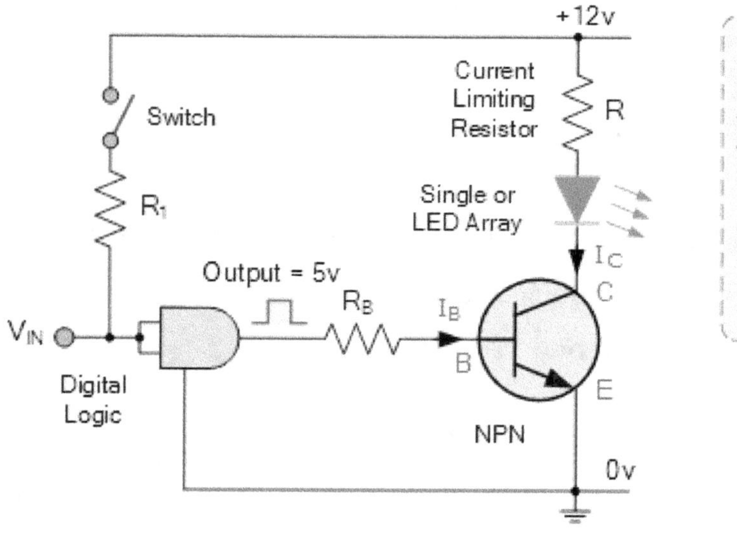

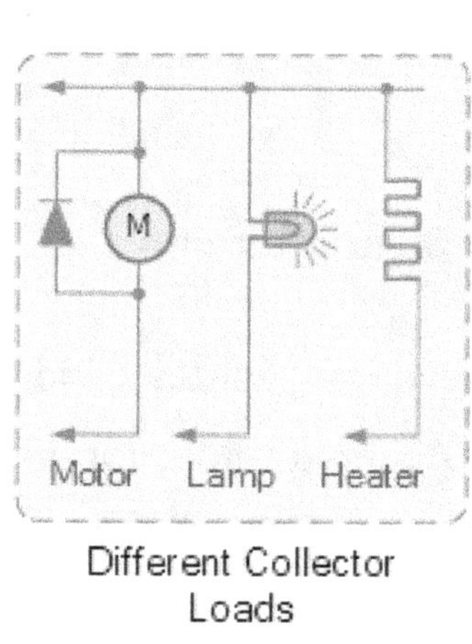

Different Collector
Loads

The base resistor, Rb is required to limit the output current from the logic gate.

PNP Transistor Switch

The PNP Transistors can also be used as a switch, the difference this time is that the load is connected to ground (0v) and the PNP transistor switches the power to it. To turn the PNP transistor operating as a switch "ON", the Base terminal is connected to ground or zero volts (LOW) as shown.

PNP Transistor Switching Circuit

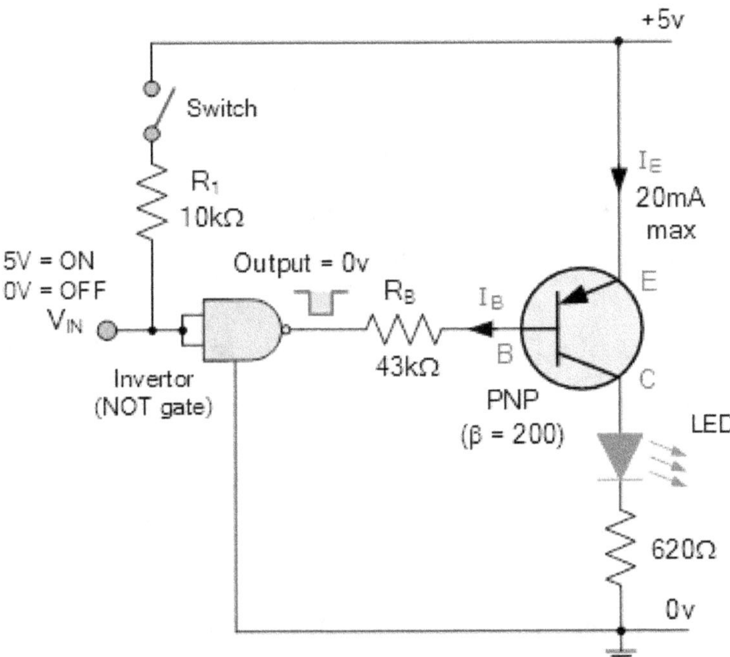

The equations for calculating the Base resistance, Collector current and voltages are exactly the same as for the previous NPN transistor switch. The difference this time is that we are switching power with a PNP transistor (sourcing current) instead of switching ground with an NPN transistor (sinking current).

Darlington Transistor Switch

The DC current gain of the bipolar transistor is often too low to directly switch the load current or voltage, so multiple switching transistors are used. Here, one small input transistor is used to switch "ON" or "OFF" a much larger current handling output transistor.

In order to maximize the signal gain, the two transistors are connected in a "Complementary Gain Compounding Configuration" or what is more commonly called a **"Darlington Configuration"** were the amplification factor is the product of the two individual transistors.

Darlington Transistors consists of two individuals bipolar NPN or PNP type transistors connected together so that the current gain of the first transistor is multiplied with that of the current gain of the second transistor to produce a device which acts like a single transistor with a very high current gain for a much smaller Base current.

The overall current gain Beta (β) or hfe value of a Darlington device is the product of the two individual gains of the transistors and is given as:

$$\beta_{TOTAL} = \beta_1 \times \beta_2$$

Therefore, Darlington Transistors with very high β values and high Collector currents are possible compared to a single transistor switch. For example, if the first input transistor has a current gain of 100 and the second switching transistor has a current gain of 50 then the total current gain will be 100 * 50 = 5000.

Here, for example, if our load current from above is 200mA, then the darlington base current is only 200mA/5000 = 40uA. A huge reduction from the previous 1mA for a single transistor.

An example of the two basic types of Darlington transistor configurations is given below.

Configurations of Darlington Transistor

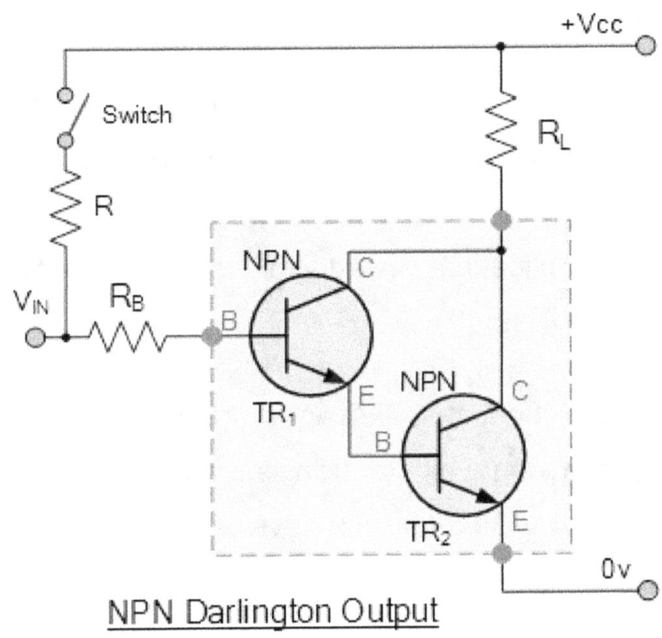

NPN Darlington Output

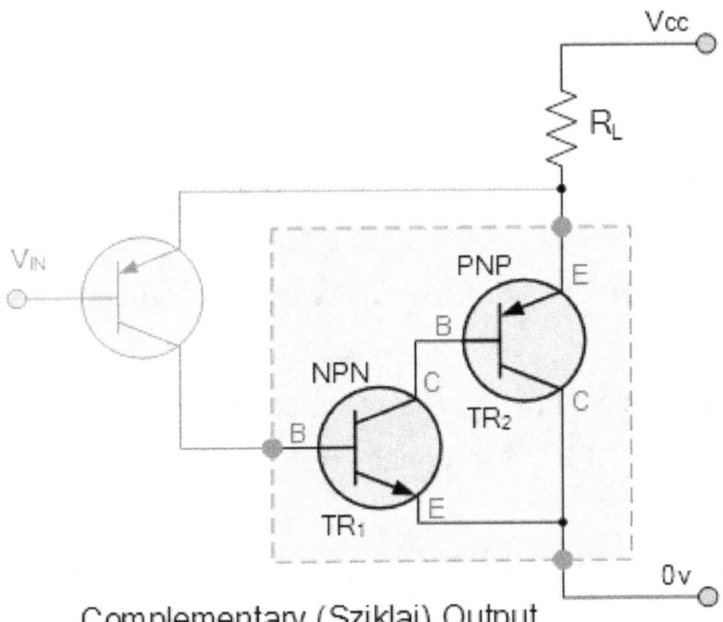

Complementary (Sziklai) Output

The Emitter of the first transistor is connected to the Base terminal of the second transistor in the above NPN Darlington transistor switch configuration, turning on the second transistor as a result of the first transistor's Emitter current becoming the second transistor's Base current. The input signal is delivered to the base of the first or "input" transistor. This transistor utilizes it to drive the larger second "output" transistors after amplifying it in the normal manner. An extremely high current gain is produced by the second transistor's second amplification of the signal.

High current gains vs single bipolar transistors are one of the primary properties of Darlington transistors. Another benefit of a "Darlington Transistor Switch" is their rapid switching speeds, which makes them perfect for use in inverter circuits, lighting circuits, and DC motor or stepper motor control applications in addition to their significantly boosted current and voltage switching capabilities.

Due to the series connection of the two PN junctions, Darlington transistors require a greater Base-Emitter input voltage (V_{BE}) than single bipolar transistors do when used as switches. This higher V_{BE} is approximately 1.4 volts for silicon devices.

Transistor as a Switch In A Brief

To summarize, the following conditions must be met when utilizing a transistor as a switch:

> ➢ Lamps, relays, and even motors can be switched and controlled using transistor switches.
> ➢ The bipolar transistor must be "fully-OFF" or "fully-ON" when being used as a switch.
> ➢ When a transistor is fully "ON," it is said to be in its saturation region.
> ➢ When a transistor is completely "OFF," it is considered to be in its cut-off zone.
> ➢ A modest Base current control a considerably greater Collector load current when utilizing the transistor as a switch.
> ➢ A "Flywheel Diode" is employed when transistors are used to switch inductive loads like relays and solenoids.
> ➢ Darlington Transistors can be used to manage enormous currents or voltages.

In N-channel and P-channel topologies, the Junction Field Effect Transistor, or JFET, is a voltage-controlled three-terminal unipolar semiconductor device. The Junction Field Effect Transistor is a unipolar device in which the electric field at a reverse biased pn-junction regulates the flow of current between its two electrodes.

We know that the transistor's output Collector current is inversely proportional to the input current flowing into its Base terminal. As a result, the bipolar transistor can switch a bigger load current using a lesser current, making it a "CURRENT" operated device (Beta model).

However, the Field Effect Transistor, or simply FET, controls the current flowing through them by using the voltage provided to their input terminal, known as the Gate, causing the output current to be proportional to the input voltage. The Field Effect Transistor is a "VOLTAGE" controlled device since its operation depends on an electric field (hence the term field effect) produced by the input Gate voltage.

Typical Field Effect Transistor

A three terminal unipolar semiconductor device called a field effect transistor, it has many traits with its bipolar transistor cousins. For instance, they can be utilized to replace their comparable bipolar junction transistors (BJT) relatives in the majority of electronic circuit applications because to their high efficiency, instantaneous operation, robustness, and affordability.

Field effect transistors are suitable for use in integrated circuits, such as the CMOS line of digital logic chips, because they can be made considerably smaller than an equivalent BJT transistor and have low power consumption and power dissipation.

NPN and PNP, which basically characterize the physical arrangement of the P-type and N-type semiconductor materials from which they are created, are the two fundamental ways that bipolar transistors are constructed, as we recall from the previous lessons.

This is also true with FETs, which can be divided into two categories: N-channel FETs and P-channel FETs, respectively. The main current carrying channel between the Drain and Source terminals of the field effect transistor, which has three terminals, is built without any PN-junctions. These terminals perform the functions of the bipolar transistor's Collector and Emitter, respectively. The "channel," which can be made of either a P-type or an N-type semiconductor material, is the current path that runs between these two terminals.

By adjusting the voltage given to the Gate, it is possible to regulate the amount of current flowing through this channel. Bipolar Transistors, as their name suggests, are "Bipolar" electronics because they function with both electrons and holes as charge carriers. On the other side, the Field Effect Transistor is a "Unipolar" device that simply relies on the conduction of electrons (N-channel) or holes (P-channel).

The input impedance (Rin) of the Field Effect Transistor is extremely high (thousands of Ohms), whereas that of the BJT is relatively low, giving it a significant advantage over its conventional bipolar transistor brethren. They are particularly sensitive to input voltage signals because of their extremely high input impedance, but this sensitivity comes at a cost because static electricity may quickly harm them. The Insulated-gate

Field Effect Transistor (IGFET), also known as the ordinary Metal Oxide Semiconductor Field Effect Transistor or MOSFET for short, and the Junction Field Effect Transistor (JFET) are the two primary varieties of field effect transistors.

The Junction Field Effect Transistor

We know that the primary current carrying path between the Emitter and Collector terminals of a bipolar junction transistor is built utilizing two PN-junctions. The Junction Field Effect Transistor (JUGFET or JFET) lacks PN-junctions in favor of a narrow, high resistivity semiconductor "Channel" made of either N-type or P-type silicon, with two ohmic electrical connections at either end that are known as the Drain and the Source, respectively.

The two basic types of junction field effect transistors (JFETs) are N-channel and P-channel devices. The channel of an N-channel JFET is doped with donor impurities, causing electrons to flow through the channel negatively (hence the name N-channel).

The P-channel JFET's channel is similarly doped with acceptor impurities, resulting in positive (hence the name P-channel) current flow across the channel in the form of holes. Since electrons are more mobile via a conductor than holes, N-channel JFETs offer higher channel conductivity (lower resistance) than their comparable P-channel versions. Because of this, N-channel JFETs are more effective conductors than P-channel JFETs.

We know that the Drain and the Source are two ohmic electrical connections located at opposite end of the channel. The Gate terminal, which can also be a P-type or N-type material and forms a PN-junction with the main channel, is a third electrical connection that exists within this channel. Below, the connections of a bipolar junction transistor and a junction field effect transistor are compared.

Comparison of Connections between a JFET and a BJT

Bipolar Transistor (BJT)   Field Effect Transistor (FET)

Emitter – (E)   >>   Source – (S)

Base – (B)   >>   Gate – (G)

Collector – (C)   >>   Drain – (D)

The symbols and basic construction for both configurations of JFETs are shown below.

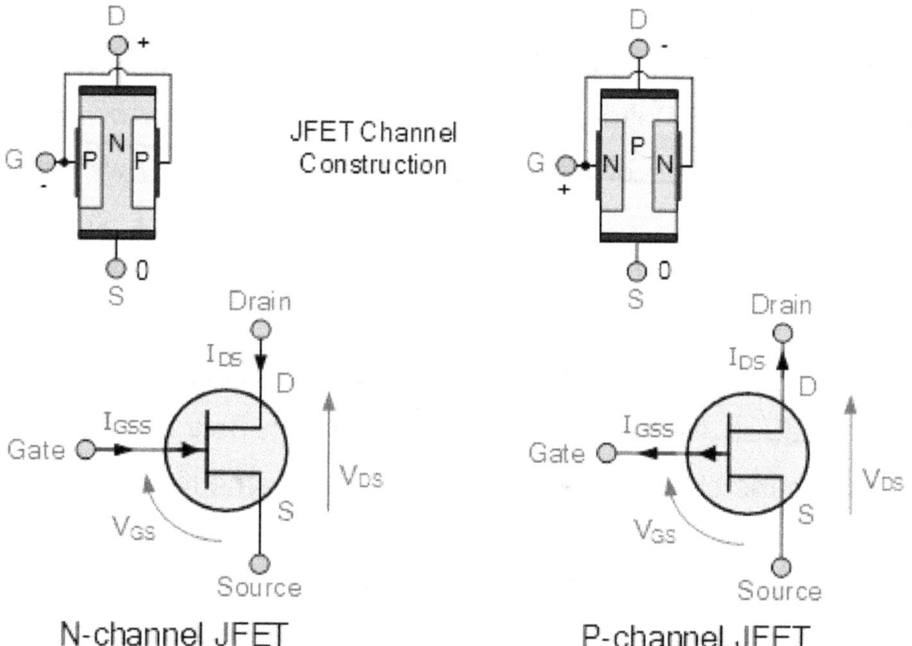

The junction field effect transistor can conduct current equally well in either direction because the semiconductor "channel" of the device is a resistive conduit through which a voltage VDS causes a current ID to flow. Due to the resistive nature of the channel, a voltage gradient develops along its length, becoming less positive as we move from the Drain terminal to the Source terminal.

As a result, the reverse bias at the PN-junction is higher at the Drain terminal and lower at the Source terminal. This bias results in the formation of a "depletion layer" within the channel, whose breadth rises with the bias. A voltage applied to the Gate terminal, which is reverse-biased, controls how much current flows through the channel between the Drain and the Source terminals.

A P-channel JFET has a positive gate voltage, whereas an N-channel JFET has a negative gate voltage. The major distinction between a JFET and a BJT device is that the Base current of a BJT is always some value larger than zero, whereas the Gate current of a JFET is essentially zero when the junction is reverse-biased.

Biasing of an N-channel Junction Field Effect Transistor

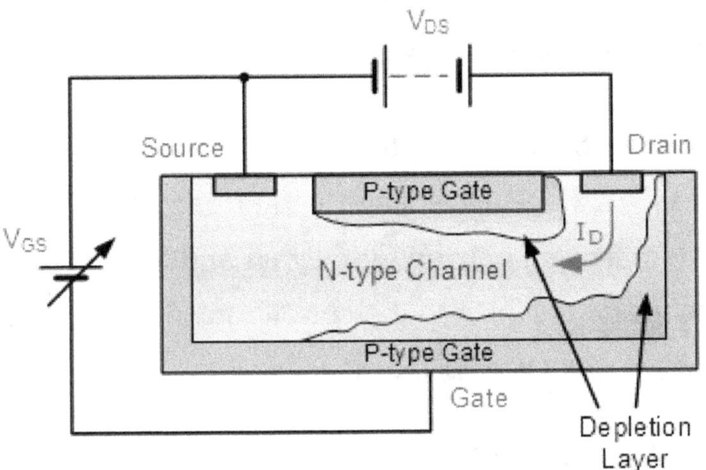

The N-type semiconductor channel shown in the cross sectional diagram above has a P-type region known as the Gate diffused into it, generating a reverse biased PN-junction. When no external voltages are applied, this junction creates the depletion region around the Gate area.

Be a result, JFETs are referred to as depletion mode devices. By reducing the effective width of the channel and hence raising the channel's overall resistance, this depletion zone creates a potential gradient around the PN-junction that is of varied thickness and restricts current flow through the channel.

The depletion region's most depleted area is between the Gate and the Drain, while the least depleted area is between the Gate and the Source, as can be seen. The JFET's channel then conducts when no bias voltage is provided (ie, the depletion region has near zero width).

Maximum saturation current (I_{DSS}) will flow through the channel from the Drain to the Source confined only by the small depletion zone surrounding the junctions with no external Gate voltage ($V_G = 0$) and a modest voltage (V_{DS}) provided between the Drain and the Source.

A kind of "squeezing" effect occurs if a tiny negative voltage ($-V_{GS}$) is now supplied to the gate because it causes the size of the depletion region to expand, decreasing the total effective area of the channel and hence reducing the current flowing through it. Therefore, increasing the breadth of the depletion region by providing a reverse bias voltage lowers the channel's conductivity by widening the depletion region.

Little current will flow into the gate connection because the PN-junction is reverse biased. The width of the channel shrinks as the Gate voltage ($-V_{GS}$) becomes more negative until there is no longer any current flowing between the Drain and the Source, at which point the FET is said to be "pinched-off" (similar to the cut-off region for a BJT). The "pinch-off voltage" is the voltage at which the channel closes (V_P).

Junction Field effect Transistor Channel Pinched-off

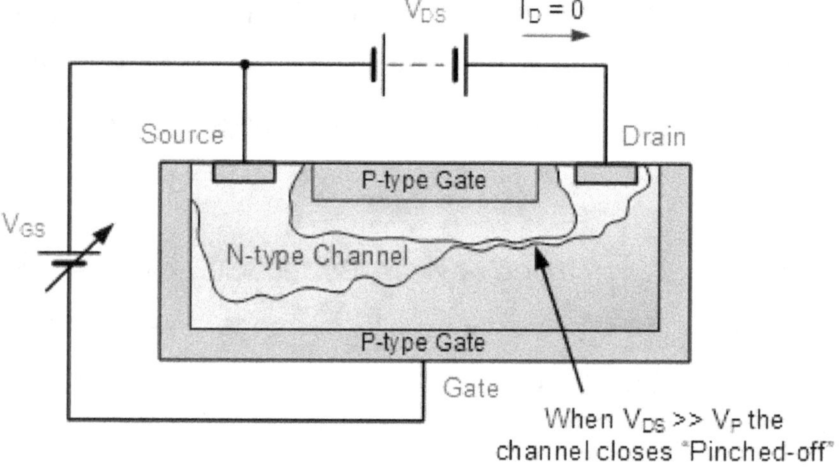

In this pinch-off region the Gate voltage, V_{GS} controls the channel current and V_{DS} has little or no effect.

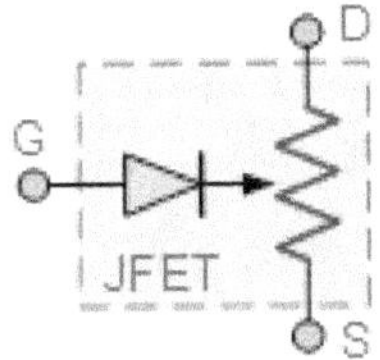

JFET Model

As a result, the FET behaves more like a voltage-controlled resistor, with a high "ON" resistance (R_{DS}) when the gate voltage is extremely negative and zero resistance when $V_{GS} = 0$. The JFET gate is always negatively biased in relation to the source when it is operating normally. The Gate voltage must always be negative because if it is, the entire channel current will flow to the Gate instead of the Source, damaging the JFET.

After that, the channel will be shut down:

- V_{DS} and No Gate Voltage (V_{GS}) are raised from zero.
- Gate control decreases adversely from zero when there is no V_{DS}.
- Varying V_{DS} and V_{GS}.

With the following exceptions, the P-channel Junction Field Effect Transistor functions identically like the N-channel transistor described above:

1) Due to holes, the channel current is positive, and
2) The biasing voltage has to have its polarity switched.

The following are the output characteristics of an N-channel JFET with the gate shorted to the source:

Output characteristic V-I curves of a typical junction FET

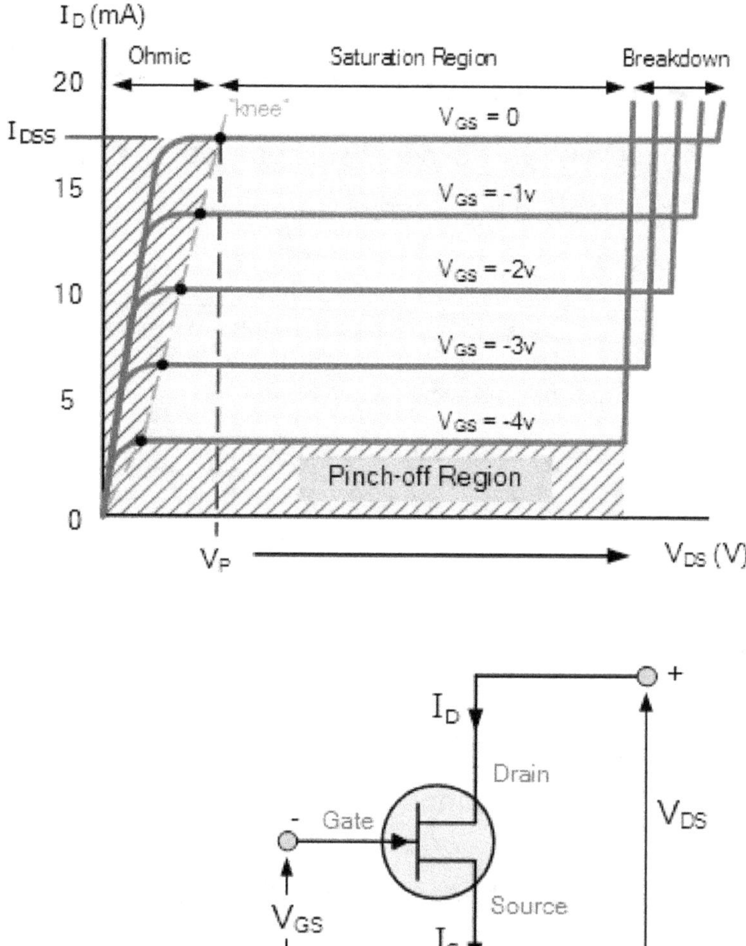

The current flowing between the Drain and the Source terminals is controlled by the voltage V_{GS} applied to the Gate. The voltage applied between the Gate and the Source is referred to as V_{GS}, whereas the voltage applied between the Drain and the Source is referred to as V_{DS}. If no current flows into the gate of a junction field effect transistor, which is a voltage-controlled device, then the source current (I_S) flowing out of the device equals the drain current flowing into it, and so ($I_D = I_S$). The four distinct zones of operation for a JFET are depicted in the example of characteristics curves above, and they are as follows:

- **Ohmic Region:** The JFET behaves like a voltage-controlled resistor in this region when $V_{GS} = 0$ because the channel's depletion layer is very thin.
- **Cut-off Region:** This is also referred to as the pinch-off region and is where the JFET behaves as an open circuit since the channel resistance is at its highest.
- **Saturation or Active Region:** During this phase, the JFET behaves like a good conductor and is controlled by the gate-source voltage (V_{GS}), with little to no influence from the drain-source voltage (V_{DS}).
- **Breakdown Region:** The JFET's resistive channel breaks down in this region, allowing an unregulated maximum current to flow. This voltage between the Drain and the Source (V_{DS}) is high enough to cause this.

The P-channel junction field effect transistor's characteristic curves are identical to those previously mentioned, with the exception that the Drain current I_D drops as the positive Gate-Source voltage, V_{GS}, increases.

When $V_{GS} = V_P$, the Drain current is zero. V_{GS} is skewed to be somewhere between V_P and 0 during normal operation. The Drain current, I_D, can thus be calculated as follows for any given bias point in the saturation or active region:

Drain current in the active region:

$$I_D = I_{DSS}\left[1 - \frac{V_{GS}}{V_P}\right]^2$$

It should be noted here that the value of the Drain current will be between zero (pinch-off) and I_{DSS} (maximum current). By knowing the Drain current I_D and the Drain-Source voltage V_{DS} the resistance of the channel (R_{DS}) is given as:

Drain-Source Channel Resistance.

$$R_{DS} = \frac{\Delta V_{DS}}{\Delta I_D} = \frac{1}{g_m}$$

Here, g_m is the "transconductance gain" since the JFET is a voltage controlled device and which represents the rate of change of the Drain current with respect to the change in Gate-Source voltage.

Modes of FET's

The field effect transistor, which has three terminals like the bipolar junction transistor and can operate in three different modes, can be connected to other components of a circuit in one of the following ways.

Configuration of Common Source (CS)

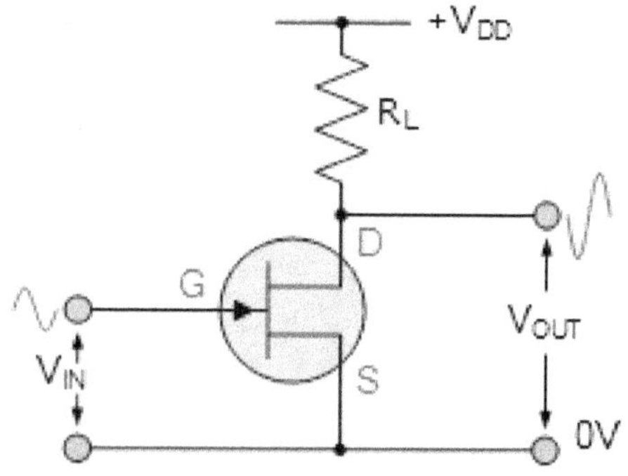

The input is applied to the Gate and its output is collected from the Drain in the common source configuration (similar to common emitter), as shown. Due to the high input impedance and effective voltage amplification of this FET operating mode, Common Source amplifiers are frequently used. In high input impedance pre-amps and stages, as well as audio frequency amplifiers, the common source form of FET connection is frequently employed. The output signal from an amplifying circuit is 180⁰ "out-of-phase" with the input.

Configuration of Common Gate (CG)

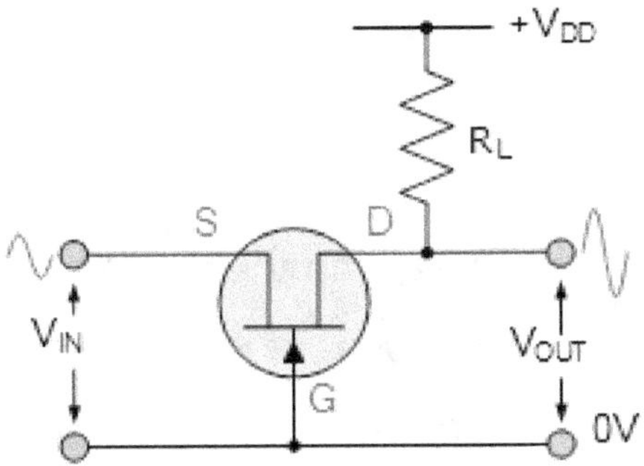

Similar to common base, the Common Gate setup applies input to the Source and draws output from the Drain with the Gate connected directly to ground (0v), as indicated. The common gate in this design has a low input impedance but a high output impedance, negating the high input impedance feature of the preceding connection. When a low input impedance needs to be matched to a high output impedance, such as in high frequency circuits or impedance matching circuits, this type of FET design can be employed. The output and the input are "in-phase."

Configuration of Common Drain (CD)

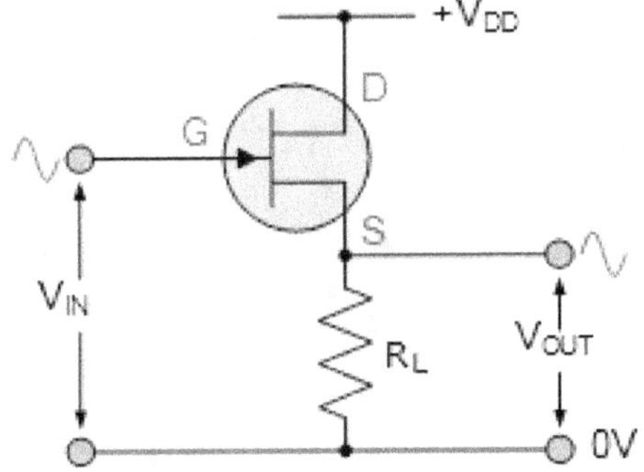

Similar to common collector, the input is applied to the Gate in a common drain arrangement, while the Gate's output is obtained from the Source. The common drain or "source follower" design is employed in buffer amplifiers because it has a near-unity voltage gain, a high input impedance, and a low output impedance.

The output signal is "in-phase" with the input signal and the voltage gain of the source follower arrangement is less than unity. Because there is no signal present at the drain connection and $+V_{DD}$ is only providing a bias, this configuration type is known as "Common Drain." The output and the input are in phase.

The Junction Field Effect Transistor Amplifier

JFETs can be utilized to create single stage class A amplifier circuits, and their common source amplifier and properties are extremely comparable to those of the bipolar junction transistor (BJT) common emitter circuit. JFET amplifiers have a high input impedance, which is controlled by the gate biasing resistive network made up of R1 and R2, as depicted, giving them an edge over BJT amplifiers.

Biasing of the Junction Field Effect Transistor Amplifier

$$V_S = I_D R_S = \frac{V_{DD}}{4}$$

$$V_S = V_G - V_{GS}$$

$$V_G = \left(\frac{R2}{R1 + R2} \right) V_{DD}$$

$$I_D = \frac{V_S}{R_S} = \frac{V_G - V_{GS}}{R_S}$$

This common source (CS) amplifier circuit is biased in class "A" mode by the voltage divider network formed by resistors R1 and R2. The voltage across the Source resistor

R_S is generally set to be about one quarter of V_{DD}, (V_{DD} /4) but can be any reasonable value.

The required Gate voltage can then be calculated from this R_S value. Since the Gate current is zero, (I_G = 0) we can set the required DC quiescent voltage by the proper selection of resistors R1 and R2.

The Junction Field Effect Transistor's ability to control the Drain current by a negative Gate potential makes it useful as a switch. For an N-channel JFET, it is crucial that the Gate voltage never be positive because this will cause the channel current to flow to the Gate instead of the Drain, damaging the JFET. The operating principles of a P-channel JFET are identical to those of an N-channel JFET, with the exception that the polarity of the voltages must be switched.

Despite having a gate terminal that is electrically isolated from the conductive channel, MOSFETs function similarly to JFETs. There is a different kind of Field Effect Transistor, in addition to the Junction Field Effect Transistor (JFET), whose Gate input is electrically isolated from the primary current carrying channel. A semiconductor device class known as an Insulated Gate Field Effect Transistor includes the MOSFET.

The Metal Oxide Semiconductor Field Effect Transistor, or MOSFET for short, is the most popular kind of insulated gate FET and is used in many various kinds of electronic circuits. The IGFET, also known as a MOSFET, is a voltage-controlled field effect transistor that differs from a JFET in that it has a "Metal Oxide" Gate electrode that is electrically isolated from the primary semiconductor, known as the n-channel or p-channel, by a very thin layer of insulating material, typically silicon dioxide, also known as glass.

This extremely thin insulated metal gate electrode can be compared to a capacitor's base plate. The input resistance of the MOSFET is so high, up in the Mega-ohms ($M\Omega$) range, due to the isolation of the regulating Gate, that it is practically unlimited.

Since the main current-carrying channel between the drain and source is electrically isolated from the gate terminal, "NO current flows into the gate," and the MOSFET functions similarly to a voltage-controlled resistor in that the input voltage determines how much current flows through the main channel between the drain and source.

Similar to the JFET, the MOSFET can also develop significant amounts of static charge due to its extremely high input resistance, which makes it vulnerable to damage if not handled properly or protected.

MOSFETs are three-terminal devices featuring a Gate, Drain, and Source, similar to the JFET tutorial before them. P-channel (PMOS) and N-channel (NMOS) MOSFETs are both readily available. The primary distinction this time is that there are now two fundamental types of MOSFETs:

- ➢ **Depletion Type:** To turn the device "OFF," the transistor needs the Gate-Source voltage (V_{GS}). A "Normally Closed" switch is analogous to a depletion mode MOSFET.
- ➢ **Enhancement Type:** For the transistor to turn the component "ON," a Gate-Source voltage (V_{GS}) is needed. A "Normally Open" switch would be comparable to an enhancement mode MOSFET.

The symbols and basic construction for both configurations of MOSFETs are shown below.

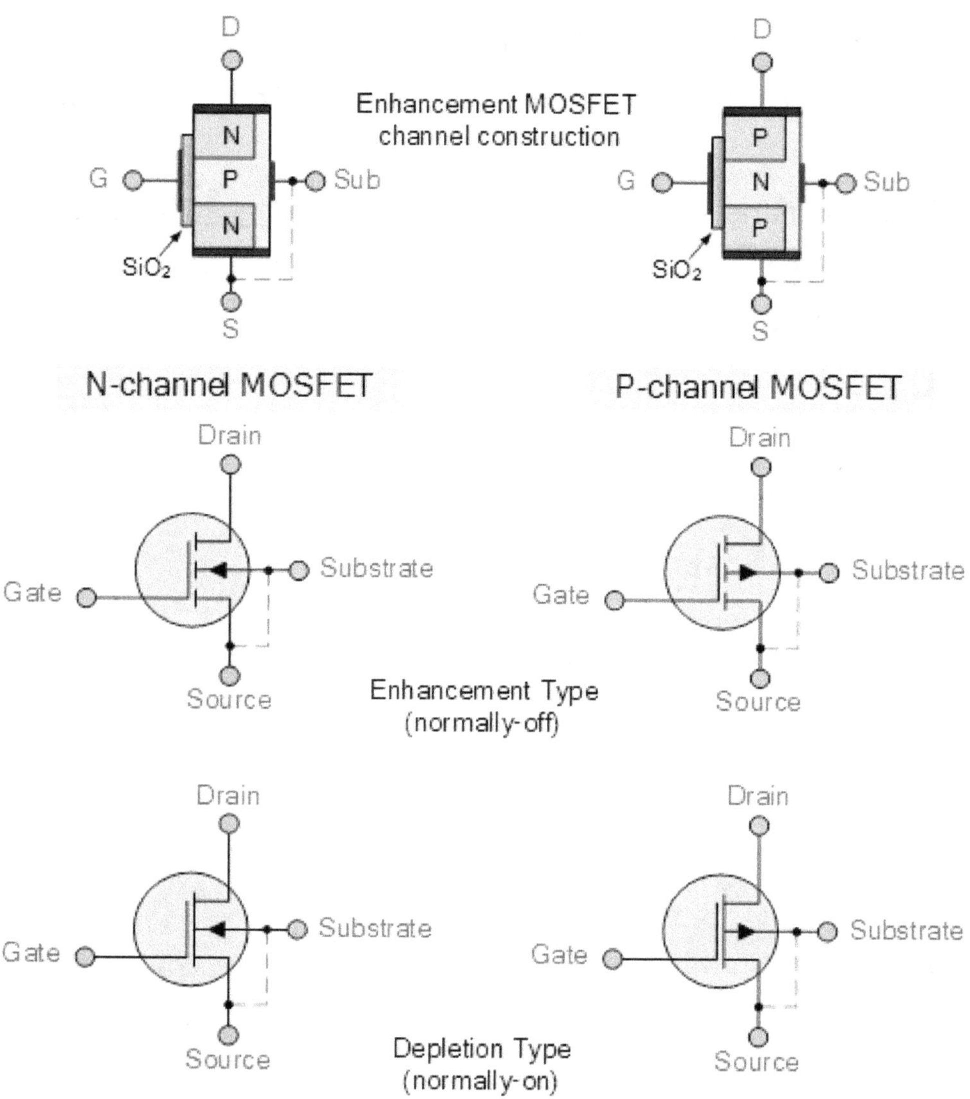

The four MOSFET symbols shown above also include a terminal labeled Substrate, which is often used for grounding the substrate rather than as an input or output connection. Through a diode junction, it is connected to the MOSFET's metal tab or body's primary semiconductive channel. This substrate lead is often internal to the source terminal of discrete type MOSFETs. When this is the case, as in the case of enhancement types, it is left out of the symbol to provide more clarity.

The semiconductive channel of the transistor is represented by the line joining the drain (D) and source (S) connections in the MOSFET symbol. As drain current can flow with zero gate biasing potential, a "Depletion" (normally-ON) type MOSFET is represented by this channel line if it is a solid, uninterrupted line.

A MOSFET of the "Enhancement" (normally-OFF) type is one in which the channel line is depicted as a dotted or broken line, as zero drain current flows with zero gate voltage. The P-type or N-type semiconductor device used in the conductive channel is indicated by the direction of the arrow pointing to this channel line.

Basic MOSFET Structure and Symbol

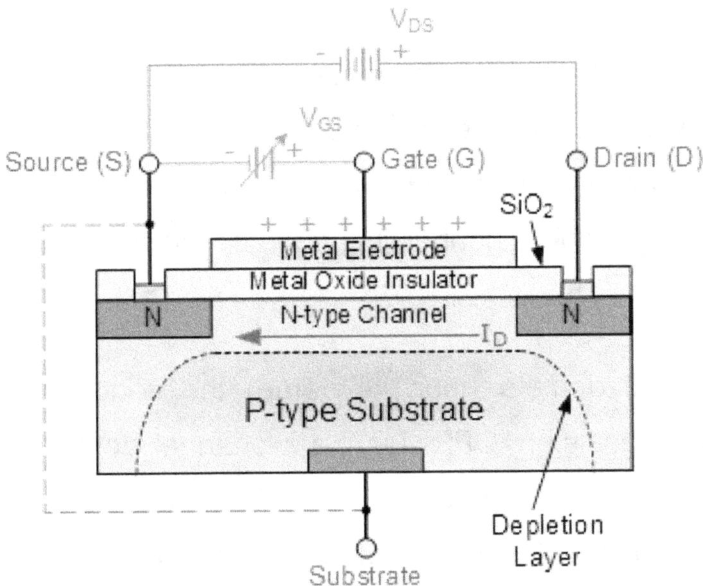

Both the Depletion and Enhancement type MOSFETs use an electrical field produced by a gate voltage to change the flow of charge carriers, electrons for n-channel or holes for P-channel, through the semiconductive drain-source channel. The Metal Oxide Semiconductor FET is built very differently from the Junction FET.

There are two tiny n-type regions right beneath the drain and source electrodes, and the gate electrode is positioned on top of a very thin insulating layer.

The gate of a junction field effect transistor, or JFET, needs to be biased in order to reverse-bias the pn-junction. No such restrictions exist when using insulated gate MOSFET devices, which makes it possible to bias the gate of a MOSFET in either polarity, positive (+ve) or negative (-ve) (-ve).

As they are generally non-conducting without bias and because MOSFETs are voltage-controlled devices, very little or no control current is required when using them as electronic switches or to create logic gates, which increases their value. The Enhancement type and the Depletion type are the two main configurations for both p-channel and n-channel MOSFETs.

Depletion-mode MOSFET

Depletion-mode MOSFETs, which are less prevalent than enhancement mode kinds, are typically turned "ON" (conducting) without the use of a gate bias voltage. As a result, the device is considered to be "normally closed" when $V_{GS} = 0$. The normally closed conductive channel is represented by a solid channel line in the circuit symbol for a depletion MOS transistor shown above.

A negative gate-source voltage ($-V_{GS}$) will deplete (thus its name) the conductive channel of its free electrons flipping the transistor "OFF" for the n-channel depletion MOS transistor. Similarly, a positive gate-source voltage ($+V_{GS}$) will exhaust the channel of its free holes, turning off a p-channel depletion MOS transistor. In other words, $+V_{GS}$ denotes greater electrons and current for an n-channel depletion mode MOSFET. While a $-V_{GS}$ denotes a reduction in electrons and current. For the p-channel varieties, the inverse is likewise accurate. The depletion mode MOSFET is hence comparable to a switch that is "usually closed."

Depletion-mode N-Channel Circuit Symbols

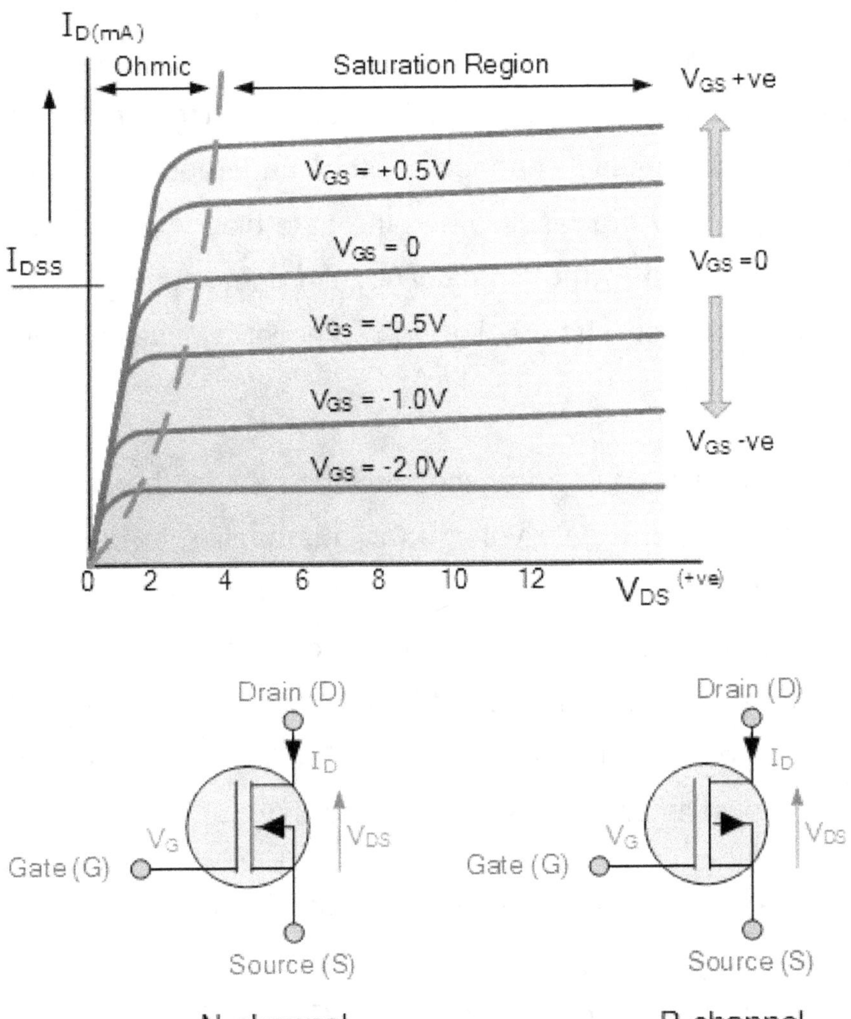

The drain-source channel of depletion-mode MOSFETs is intrinsically conductive due to the electrons and holes that are already present in the n-type or p-type channel, similar to their JFET transistor counterparts. With no Gate bias, this channel doping creates a low resistance conducting path between the Drain and Source.

MOSFET in enhancement mode

The depletion-mode type is the opposite of the more widely used enhancement-mode MOSFET, or eMOSFET. Here, the conducting channel is either underdoped or only mildly doped, rendering it nonconductive. As a result, when the gate bias voltage, V_{GS}, is equal to zero, the device is generally "OFF" (not conducting). An enhancement MOS transistor's circuit symbol employs a broken channel line to represent a usually open, non-conducting channel.

The n-channel enhancement MOS transistor is a transconductance device because a drain current will only flow through it when a gate voltage (V_{GS}) is applied to the gate terminal that is larger than the voltage level at which conductance occurs (V_{TH}). When a positive gate voltage (+ve) is applied to an n-type eMOSFET, more electrons are drawn to the oxide layer surrounding the gate, increasing the thickness of the channel and allowing more current to pass. Since the application of a gate voltage improves the channel, this form of transistor is known as an enhancement mode device.

The channel resistance will decrease even further when the positive gate voltage is raised, increasing the drain current, or I_D, through the channel. To put it another way, a zero or $-V_{GS}$ switches an n-channel enhancement mode MOSFET "OFF," whereas a $+V_{GS}$ turns it "ON." As a result, the enhancement-mode MOSFET is comparable to a switch that is "usually open."

For the p-channel enhancement MOS transistor, the opposite is true. The device is "OFF" and the channel is open when $V_{GS} = 0$. The p-type eMOSFET is turned "ON" by applying a negative (-ve) gate voltage, which increases the channel conductivity. In the case of a p-channel enhancement mode MOSFET, $+V_{GS}$ and $-V_{GS}$ respectively turn the transistor "ON" and "OFF."

Enhancement-mode N-Channel Circuit Symbols

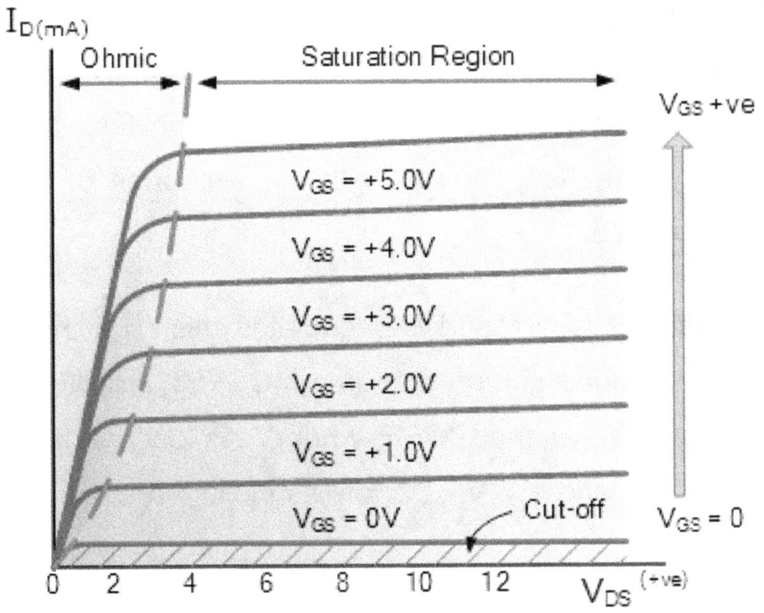

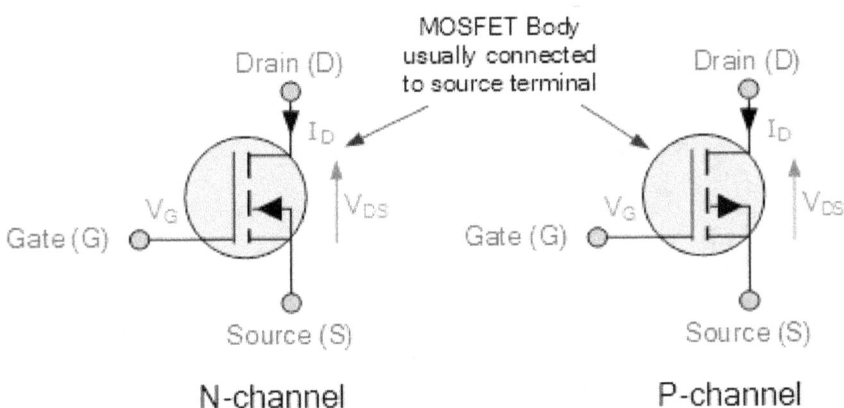

Due to its low "ON" resistance, highly high "OFF" resistance, and endlessly high input resistance because of their isolated gate, enhancement-mode MOSFETs make great electronic switches. Enhancement-mode MOSFETs are utilized in integrated circuits to provide power switching circuits with PMOS (P-channel) and NMOS (N-channel) gates and CMOS type logic gates. The term "CMOS" actually stands for "Complementary MOS," which means that the logic device's design incorporates both PMOS and NMOS.

The MOSFET Amplifier

The most common MOSFET amplifier design is the enhancement mode n-channel common source amplifier, which may be used to create single stage class "A" amplifier circuits similarly to the preceding Junction Field Effect transistor. JFET amplifiers and depletion mode MOSFET amplifiers are essentially similar, with the exception of the MOSFET's significantly higher input impedance.

The gate biasing resistive network made up of R1 and R2 regulates this high input impedance. Additionally, the enhancement mode common source MOSFET amplifier's output signal is inverted since the transistor is switched "OFF" when V_G is low and high when V_D (V_{out}). The transistor is turned "ON" and V_D (V_{out}) is low as indicated when V_G is high.

Enhancement-mode N-Channel Amplifier

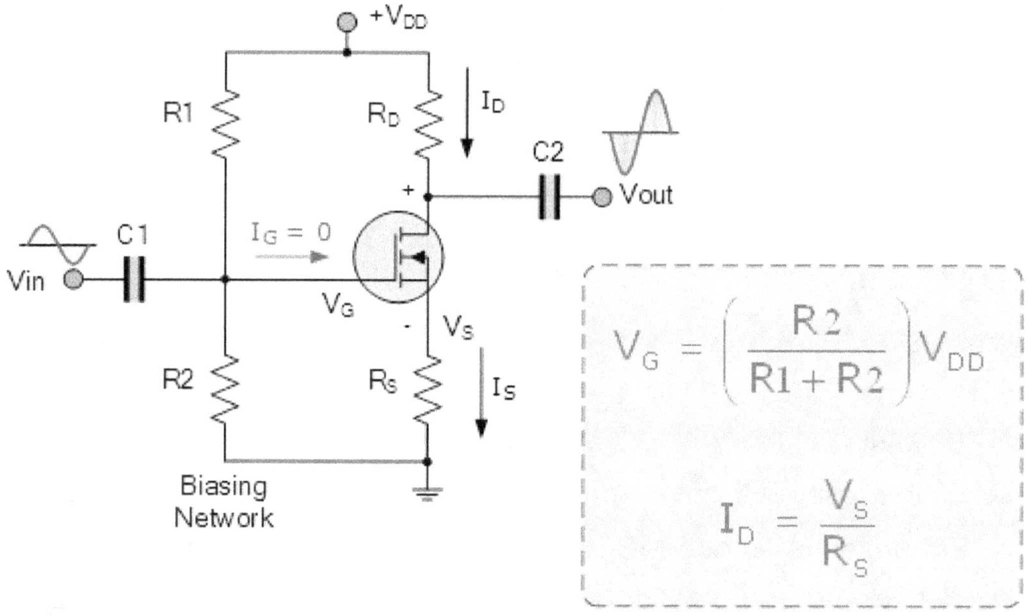

$$V_G = \left(\frac{R2}{R1 + R2} \right) V_{DD}$$

$$I_D = \frac{V_S}{R_S}$$

The DC biasing of this common source (CS) MOSFET amplifier circuit is virtually identical to the JFET amplifier. The MOSFET circuit is biased in class A mode by the voltage divider network formed by resistors R1 and R2. The AC input resistance is given as $R_{IN} = R_G = 1M\Omega$.

By applying a tiny signal voltage, metal oxide semiconductor field effect transistors—three terminal active devices constructed from various semiconductor materials—can function as either an insulator or a conductor. The MOSFET can perform two basic tasks thanks to its ability to transition between these two states: "switching" (for use in digital electronics) or "amplification" (analogue electronics). The following three zones can be operated in by MOSFETs:

1. **Cut-off Region:** with $V_{GS} < V_{threshold}$ the gate-source voltage is much lower than the transistors threshold voltage so the MOSFET transistor is switched "fully-OFF" thus, $I_D = 0$, with the transistor acting like an open switch regardless of the value of V_{DS}.
2. **Linear (Ohmic) Region:** with $V_{GS} > V_{threshold}$ and $V_{DS} < V_{GS}$ the transistor is in its constant resistance region behaving as a voltage-controlled resistance whose resistive value is determined by the gate voltage, V_{GS} level.
3. **Saturation Region:** with $V_{GS} > V_{threshold}$ and $V_{DS} > V_{GS}$ the transistor is in its constant current region and is therefore "fully-ON". The Drain current I_D = Maximum with the transistor acting as a closed switch.

MOSFET In A Brief

The input gate resistance of the Metal Oxide Semiconductor Field Effect Transistor, or MOSFET for short, is very high, and the gate voltage regulates the amount of current that flows through the channel between the source and drain. If not handled or protected correctly, MOSFETs can be readily destroyed by static electricity due to their high input impedance and gain.

Due to their low power consumption, MOSFETs are perfect for use as common-source amplifiers or electronic switches. Metal oxide semiconductor field effect transistors are commonly used in microprocessors, memories, calculators, and logic CMOS gates, among other electronic devices.

Also take note of the fact that the symbol contains a dotted or broken line indicating an enhancement type that is ordinarily "OFF," demonstrating that "NO" current can flow through the channel when zero gate-source voltage VGS is supplied. A typically "ON" Depletion type is indicated by a line that runs continuously through the entire symbol, demonstrating that current "CAN" flow through the channel with zero gate voltage. The only difference between the two symbols for p-channel types is that the arrow points outwards. The switching table that follows summarizes this.

MOSFET type	V_{GS} = +ve	V_{GS} = 0	V_{GS} = -ve
N-Channel Depletion	ON	ON	OFF
N-Channel Enhancement	ON	OFF	OFF
P-Channel Depletion	OFF	ON	ON
P-Channel Enhancement	OFF	OFF	ON

For n-type enhancement type MOSFETs, this means that the transistor is "ON" with a positive gate voltage and "OFF" with a zero-gate voltage. A negative gate voltage will "ON" a p-channel enhancement type MOSFET, while a zero-gate voltage will "OFF" the transistor. The device's threshold voltage, or V_{TH}, determines the voltage at which the MOSFET begins to flow current across the channel.

In many applications, field effect transistor switches are used to turn a DC current "ON" or "OFF" for devices like LEDs, which only need a few milliamps at low DC voltages, or motors, which need larger currents at higher voltages.

Due to the fact that they function between their cut-off and saturation regions, MOSFETs make excellent electronic switches for managing loads and in CMOS digital circuits. The N-channel, Enhancement-mode MOSFET (e-MOSFET), as we saw previously, operates with a positive input voltage and has an incredibly high input resistance (nearly infinite), making it possible to use the MOSFET as a switch when interfaced with almost any logic gate or driver capable of producing a positive output.

We also discovered that, because to the extremely high input (Gate) resistance, we can securely parallel a variety of MOSFETS till we reach the requisite current handling capacity. We can switch high currents or high voltage loads by connecting a number of MOSFETS in parallel, but doing so is expensive and wasteful of space on the circuit board and components.

Power Field Effect Transistors, sometimes known as Power FETs, were created to address this issue. We now understand that there are two key distinctions between field effect transistors: JFETs only support depletion-mode, whereas MOSFETs support both enhancement- and depletion-mode.

As these transistors need a positive gate voltage to turn "ON" and a zero gate voltage to turn "OFF," they can be simply understood as switches and are also simple to interface with logic gates. In this lesson, we will look at using the enhancement-mode MOSFET as a switch.

The I-V characteristic curves of the enhancement-mode MOSFET, or e-MOSFET, shown below, are the best representations of how the device functions. The MOSFET conducts almost little current when the input voltage (V_{IN}) to the transistor's gate is zero, and the output voltage (V_{OUT}) is equal to the supply voltage (V_{DD}). Since it is operating within its "cut-off" area, the MOSFET is "OFF."

MOSFET Characteristics Curves

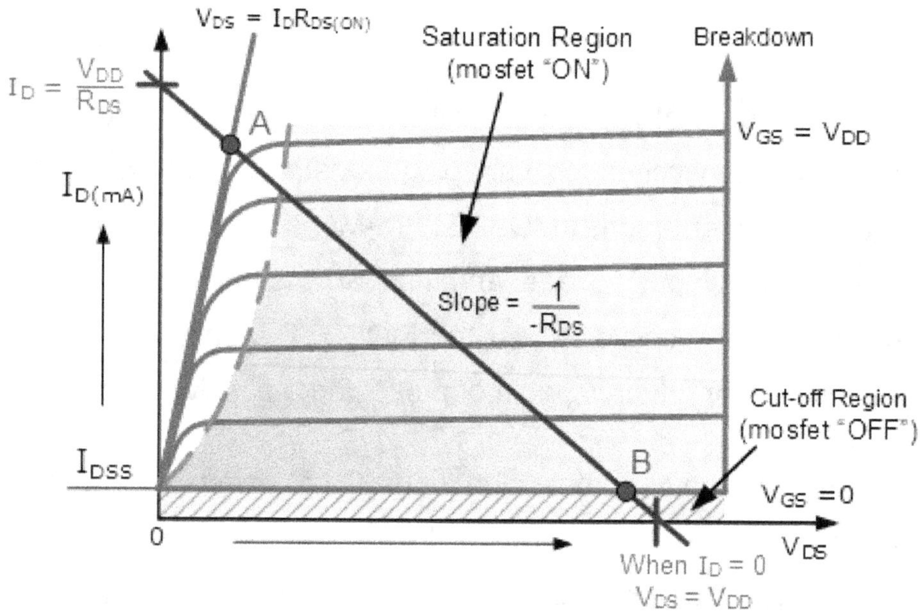

The V-I transfer curves above can be used to calculate the minimal ON-state gate voltage necessary to guarantee that the MOSFET remains "ON" while carrying the chosen drain current. The MOSFET Q-point shifts to point A along the load line when V_{IN} is HIGH or equal to V_{DD}.

A decrease in the channel resistance causes the drain current I_D to reach its maximum value. I_D turns into a constant value that is only reliant on V_{GS} and independent of V_{DD}. As a result, the transistor acts like a closed switch, but because of its $R_{DS(on)}$ value, the channel ON-resistance does not completely decrease to zero, just becoming very small. The MOSFET Q-point shifts from point A to point B along the load line when V_{IN} is LOW or lowered to zero.

Because of the high channel resistance, the transistor behaves like an open circuit and does not allow any current to flow through the channel. As a result, if the MOSFET's gate voltage alternates between HIGH and LOW, it will serve as a "single-pole single-throw" (SPST) solid state switch, and this behavior is described as:

1. Cut-off Region

Here the operating conditions of the transistor are zero input gate voltage (V_{IN}), zero drain current I_D and output voltage $V_{DS} = V_{DD}$. Therefore for an enhancement type MOSFET the conductive channel is closed and the device is switched "OFF".

Cut-off Characteristics

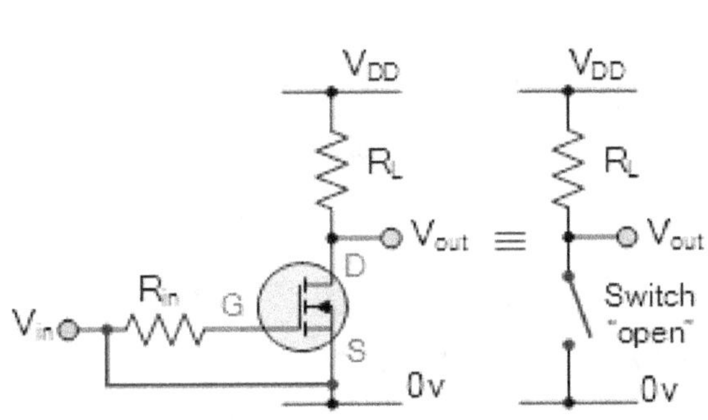

- The input and Gate are grounded (0V)
- Gate-source voltage less than threshold voltage $V_{GS} < V_{TH}$
- MOSFET is "OFF" (Cut-off region)
- No Drain current flows (I_D = 0 Amps)
- $V_{OUT} = V_{DS} = V_{DD} = $ "1"
- MOSFET operates as an "open switch"

We can define the cut-off region or "OFF mode" when using an e-MOSFET as a switch as being, gate voltage, $V_{GS} < V_{TH}$ thus I_D = 0. For a P-channel enhancement MOSFET, the Gate potential must be more positive with respect to the Source.

2. Saturation Region

In the saturation or linear region, the transistor will be biased so that the maximum amount of gate voltage is applied to the device which results in the channel resistance $R_{DS(on)}$ being as small as possible with maximum drain current flowing through the MOSFET switch. Therefore for the enhancement type MOSFET the conductive channel is open and the device is switched "ON".

Saturation Characteristics

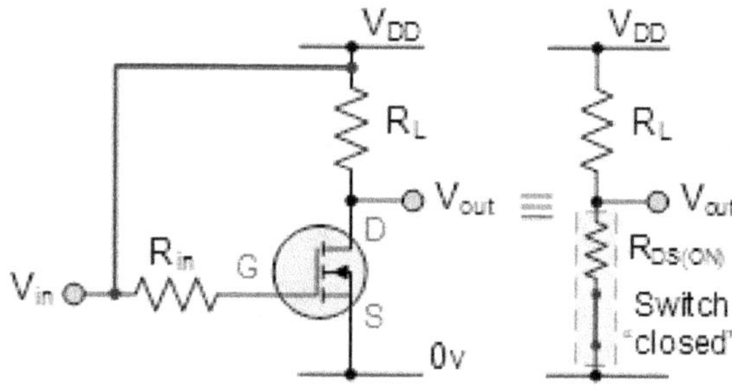

- The input and Gate are connected to V_{DD}
- Gate-source voltage is much greater than threshold voltage $V_{GS} > V_{TH}$
- MOSFET is "ON" (saturation region)
- Max Drain current flows ($I_D = V_{DD} / R_L$)
- $V_{DS} = 0V$ (ideal saturation)
- Min channel resistance $R_{DS(on)} < 0.1\Omega$
- $V_{OUT} = V_{DS} \cong 0.2V$ due to $R_{DS(on)}$
- MOSFET operates as a low resistance "closed switch"

We can define the saturation region or "ON mode" when using an e-MOSFET as a switch as gate-source voltage, $V_{GS} > V_{TH}$ thus I_D = Maximum. For a P-channel enhancement MOSFET, the Gate potential must be more negative with respect to the Source.

By applying a suitable drive voltage to the gate of an FET, the resistance of the drain-source channel, $R_{DS(on)}$ can be varied from an "OFF-resistance" of many hundreds of $k\Omega$, effectively an open circuit, to an "ON-resistance" of less than 1Ω, effectively acting as a short circuit.

When using the MOSFET as a switch we can drive the MOSFET to turn "ON" faster or slower, or pass high or low currents. This ability to turn the power MOSFET "ON" and "OFF" allows the device to be used as a very efficient switch with switching speeds much faster than standard bipolar junction transistors.

An example of using the MOSFET as a switch

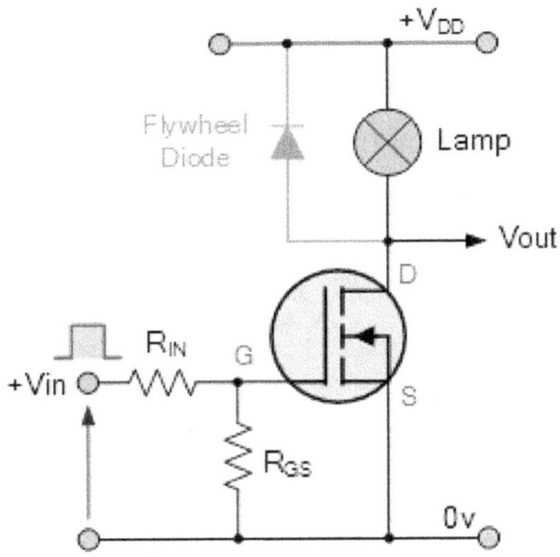

In the above circuit arrangement, an Enhancement-mode N-channel MOSFET is being used to switch a simple lamp "ON" and "OFF" (could also be an LED).

The gate input voltage V_{GS} is taken to an appropriate positive voltage level to turn the device and therefore the lamp load either "ON", (V_{GS} = +ve) or at a zero voltage level that turns the device "OFF", (V_{GS} = 0V).

If the resistive load of the lamp was to be replaced by an inductive load such as a coil, solenoid or relay a "flywheel diode" would be required in parallel with the load to protect the MOSFET from any self-generated back-emf.

Above shows a very simple circuit for switching a resistive load such as a lamp or LED. But when using power MOSFETs to switch either inductive or capacitive loads some form of protection is required to prevent the MOSFET device from becoming damaged. Driving an inductive load has the opposite effect from driving a capacitive load.

For example, a capacitor without an electrical charge is a short circuit, resulting in a high "inrush" of current and when we remove the voltage from an inductive load we have a large reverse voltage build up as the magnetic field collapses, resulting in an induced back-emf in the windings of the inductor.

We can summarize the switching characteristics of both the N-channel and P-channel type MOSFET within the following table.

MOSFET Type	$V_{GS} \ll 0$	$V_{GS} = 0$	$V_{GS} \gg 0$
N-channel Enhancement	OFF	OFF	ON
N-channel Depletion	OFF	ON	ON
P-channel Enhancement	ON	OFF	OFF
P-channel Depletion	ON	ON	OFF

It should be noted here that unlike the N-channel MOSFET whose gate terminal must be made more positive (attracting electrons) than the source to allow current to flow through the channel, the conduction through the P-channel MOSFET is due to the flow of holes. That is the gate terminal of a P-channel MOSFET must be made more negative than the source and will only stop conducting (cut-off) until the gate is more positive than the source.

Therefore, for the enhancement type power MOSFET to operate as an analogue switching device, it requires to be switched between its "Cut-off Region" where: $V_{GS} = 0V$ (or $V_{GS} = -ve$) and its "Saturation Region" where: $V_{GS(on)} = +ve$. The power dissipated in the MOSFET (P_D) depends upon the current flowing through the channel I_D at saturation and also the "ON-resistance" of the channel given as $R_{DS(on)}$. For example.

MOSFET as a Switch Example-1

Let's consider that the lamp is rated at 6v, 24W and is fully "ON", the standard MOSFET has a channel on-resistance ($R_{DS(on)}$) value of 0.1ohms. Determine the power dissipated in the MOSFET switching device. The current flowing through the lamp is calculated as:

$$P = V \times I_D$$

$$\therefore I_D = \frac{P}{V} = \frac{24}{6} = 4.0\,\text{amps}$$

Then the power dissipated in the MOSFET will be given as:

$$P = I^2.R$$

$$P_D = I_D^2 \times R_{DS}$$

$$\therefore P_D = 4^2 \times 0.1 = 1.6\,\text{watts}$$

When using the MOSFET as a switch to control DC motors or electrical loads with high inrush currents the "ON" Channel resistance ($R_{DS(on)}$) between the drain and the source is very important. For example, MOSFETs that control DC motors, are subjected to a high in-rush current when the motor first begins to rotate, because the motors starting current is only limited by the very low resistance value of the motors windings.

Since the basic power relationship is: $P = I^2R$, then a high $R_{DS(on)}$ channel resistance value would simply result in large amounts of power being dissipated and wasted within the MOSFET itself resulting in an excessive temperature rise, which if not controlled could result in the MOSFET becoming very hot and damaged due to a thermal overload.

A lower $R_{DS(on)}$ value for the channel resistance is also a desirable parameter as it helps to reduce the channels effective saturation voltage ($V_{DS(sat)} = I_D{}^*R_{DS(on)}$) across the MOSFET and will therefore operate at a cooler temperature. Power MOSFETs generally have a $R_{DS(on)}$ value of less than 0.01Ω which allows them to run cooler, extending their operational life span.

One of the main limitations when using a MOSFET as a switching device is the maximum drain current it can handle. So the $R_{DS(on)}$ parameter is an important guide to the switching efficiency of the MOSFET and is simply given as the ratio of V_{DS} / I_D when the transistor is switched "ON".

When using a MOSFET or any type of field effect transistor for that matter as a solid-state switching device it is always advisable to select ones that have a very low $R_{DS(on)}$ value or at least mount them onto a suitable heatsink to help reduce any thermal runaway and damage. Power MOSFETs used as a switch generally have surge-current protection built into their design, but for high-current applications the bipolar junction transistor is a better choice.

Power MOSFET Motor Control

Due to the extremely high input or gate resistance that the MOSFET has, its very fast switching speeds and the ease at which they can be driven makes them ideal to interface with op-amps or standard logic gates. However, care must be taken to ensure that the gate-source input voltage is correctly chosen because when using the **MOSFET as a switch** the device must obtain a low $R_{DS(on)}$ channel resistance in proportion to this input gate voltage.

Low threshold type power MOSFETs may not switch "ON" until a least 3V or 4V has been applied to its gate and if the output from the logic gate is only +5V logic it may be insufficient to fully drive the MOSFET into saturation. Using lower threshold MOSFETs designed for interfacing with TTL and CMOS logic gates that have thresholds as low as 1.5V to 2.0V are available.

Power MOSFETs can be used to drive computer logic or pulse-width modulation (PWM) type controllers to control the motion of DC motors or brushless stepper motors. Since a DC motor has a strong beginning torque and an armature current that is proportionate to it, MOSFET switches and a PWM can be employed as excellent speed controllers to ensure that the motor runs smoothly and quietly.

Simple Power MOSFET Motor Controller

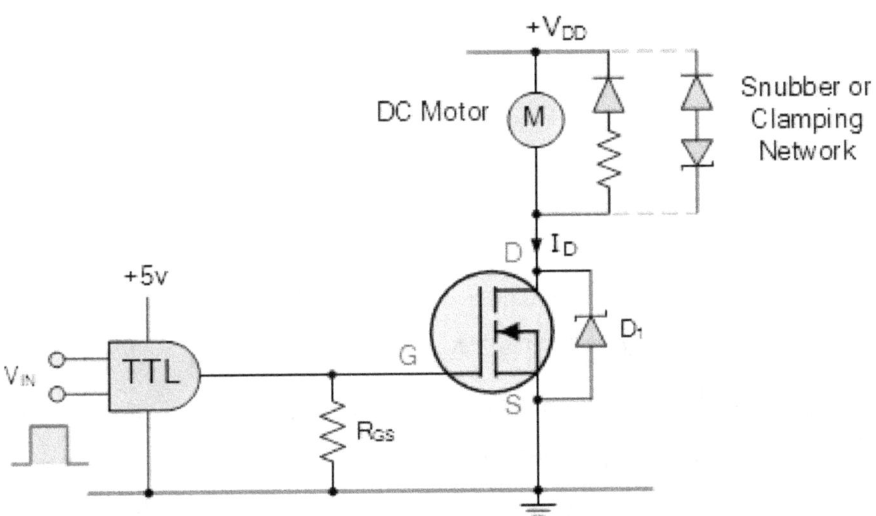

A simple flywheel diode is placed across the inductive load of the motor because the load is inductive, dissipating any back emf produced by the motor when the MOSFET turns it "OFF." Faster switching and greater control of the peak reverse voltage and drop-out time can also be achieved by using a clamping network made up of a zener diode connected in series with the diode.

For added security an additional silicon or zener diode D_1 can also be placed across the channel of a MOSFET switch when using inductive loads, such as motors, relays, solenoids, etc, for suppressing over voltage switching transients and noise giving extra protection to the MOSFET switch if required. Resistor R_{GS} is used as a pull-down resistor to help pull the TTL output voltage down to 0V when the MOSFET is switched "OFF".

P-channel MOSFET as a Switch

The N-channel MOSFET has been viewed up to this point as a switch that is positioned between the load and the ground. Additionally, the gate drive or switching signal of the MOSFET may now be referenced to ground (low-side switching).

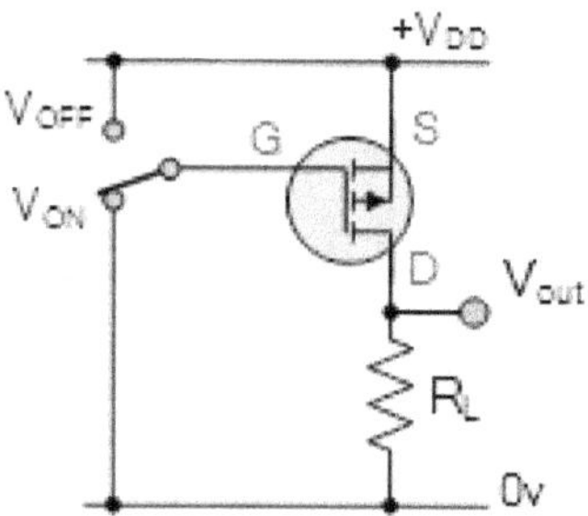

P-channel MOSFET Switch

However, in other situations, where the load is directly linked to ground, we must employ P-channel enhancement-mode MOSFETs. As with PNP transistors, the MOSFET switch in this case is connected between the load and the positive supply rail (high-side switching). The traditional direction of drain current flow in a P-channel device is downward, so to turn the transistor "ON," a downward gate-source voltage is applied.

The P-channel MOSFET is "upside down," with its source terminal connected to the positive supply $+V_{DD}$, enabling this. The MOSFET therefore turns "ON" when the switch is in the LOW position and "OFF" when the switch is in the HIGH position. It is possible to create a complementary or CMOS switching device, as shown across a dual supply, by connecting a P-channel enhancement mode MOSFET switch upside down in series with an N-channel enhancement mode MOSFET.

Complementary MOSFET as a Switch Motor Controller

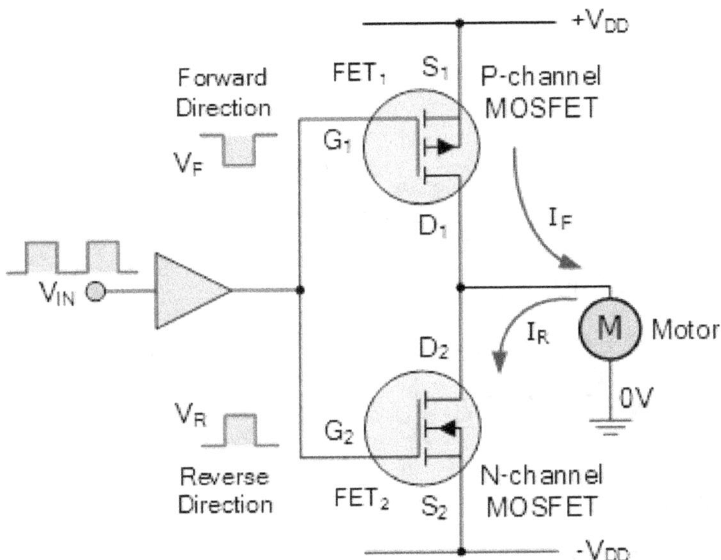

The motor is connected between the common drain connection and ground reference while the two MOSFETs are set up to form a bi-directional switch from a dual supply. The P-channel MOSFET is switched ON when the input is LOW because the gate-source junction is negatively biased, which causes the motor to rotate in one direction. To drive the motor, just the positive $+V_{DD}$ supply rail is used.

The N-channel device switches ON when the input is HIGH because its gate-source junction is positively biased, while the P-channel device switches OFF. Due to the motor's terminal voltage being supplied by the negative $-V_{DD}$ supply rail now instead of the positive $-V_{DD}$ supply rail, the motor now rotates in the other direction.

The P-channel MOSFET is then utilized for "high-side switching" to switch the positive supply to the motor in the forward direction while "low-side switching" is used to switch the negative supply to the motor in the reverse direction (low-side switching).

The two MOSFETs can be driven in a variety of ways for a wide range of applications. A single gate drive I_C, as shown, can drive both P-channel and N-channel devices. Fast switching devices must allow for some time to pass between them turning "OFF" and the other turning "ON" in order to prevent cross-conduction, which occurs when both MOSFETS conduct simultaneously across the dual supply's two polarities. Driving each MOSFETS gate separately is one technique to solve this issue. When both MOSFETS are "OFF," this results in a third option, which tells the motor to "STOP."

MOSFET as a Switch Motor Control Table

MOSFET 1	MOSFET 2	Motor Function
OFF	OFF	Motor Stopped (OFF)
ON	OFF	Motor Rotates Forward
OFF	ON	Motor Rotates Reverse
ON	ON	NOT ALLOWED

It should be noted here that no other input combinations are permitted at the same time as doing so may result in a shorted power supply as both MOSFETS, FET_1, and FET_2 may be turned "ON" at the same time, causing a fuse to blow.

For a given base current, the Darlington Transistor arrangement of two bipolar transistors allows enhanced current switching. The Darlington Transistor, which bears Sidney Darlington's name, is a unique configuration made up of two conventional NPN or PNP bipolar junction transistors (BJTs) coupled in series.

In order to create a more sensitive transistor with a significantly higher current gain that is beneficial in applications where current amplification or switching is required, the emitter of one transistor is connected to the base of the other.

A single device commercially manufactured in a single package with the standard: Base, Emitter, and Collector connecting leads can be used to create a pair of Darlington transistors. These transistors are available in a wide range of case designs and voltage (and current) ratings in both NPN and PNP versions. The bipolar junction transistor (BJT) can be designed to function as an ON-OFF switch as depicted, as we saw in our Transistor as a Switch tutorial, in addition to being utilized as an amplifier.

Bipolar Transistor as a Switch

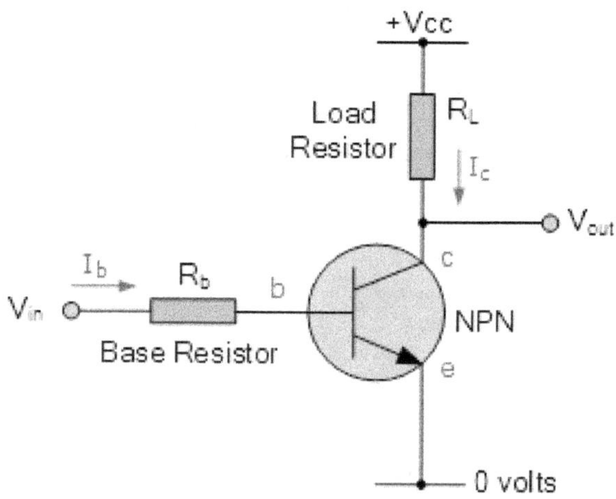

When the base terminal of the NPN transistor is grounded (0 volts), zero current flows into the base therfore Ib = 0. As the base terminal is grounded, no current flows from the collector to the emitter terminals therefore the non-conducting NPN transistor is switched "OFF" (cut-off). If we now forward biased the base terminal with respect to the emitter by using a voltage source greater than 0.7 volts, transistor action occurs causing in a much larger current to flow through the transistor between its collector and emitter terminals. The transistor is now said to be switched "ON" (conducting). If we operate the transistor between these two modes of cut-off and conduction, the transistor can be made to operate as an electronic switch.

However, the transistors base terminal needs to be switched between zero and some positive value much greater than 0.7 volts for the transistor to fully conduct. A higher voltage causes an increased base current, Ib to flows into the device resulting in collector current Ic becoming large while the voltage drop across the colletor and emitter terminals, Vce becomes smaller. Then we can see that a smaller current flowing into the base terminal can cause a much larger current to flow between the collector and the emitter.

The ratio of collector current to base current (β) is known as the *current gain* of the transistor. A typical value of β for a standard bipolar transistor may be in the range of 50 to 200 and varies even between transistors of the same part number. In some cases where the current gain of a single transistor is too low to directly drive a load, one way to increase the gain is to use a Darlington pair.

A "Darlington pair" or "super-alpha circuit," also known as a configuration of Darlington transistors, is made up of two NPN or PNP transistors coupled in such a way that the base current of the first transistor, TR1, becomes the emitter current of the second transistor, TR2. Then, as shown below, transistor TR1 is connected as an emitter follower and transistor TR2 is connected as a common emitter amplifier. Also take note that the collector currents of the master switching transistor TR2 and the slave or control transistor TR1 are "in-phase" in this Darlington pair setup.

Configuration of Basic Darlington Transistor

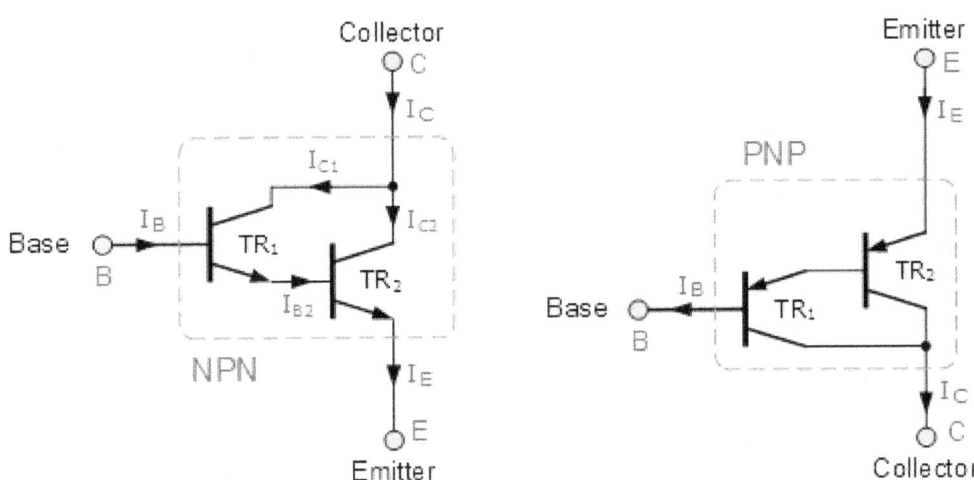

Utilizing the NPN Darlington pair as the example, the collectors of two transistors are connected together, and the emitter of TR_1 drives the base of TR_2. This configuration achieves β multiplication because for a Base current i_b, the collector current is $β*i_b$ where the current gain is greater than one, or unity and this is defined as:

$$I_C = I_{C1} + I_{C2}$$

$$I_C = \beta_1.I_B + \beta_2.I_{B2}$$

But the base current, I_{B2} is equal to transistor TR1 emitter current, I_{E1} as the emitter of TR1 is connected to the base of TR2. Therefore:

$$I_{B2} = I_{E1} = I_{C1} + I_B = \beta_1.I_B + I_B = (\beta_1 + 1).I_B$$

Then substituting in the first equation:

$$I_C = \beta_1.I_B + \beta_2.(\beta_1 + 1).I_B$$

$$I_C = \beta_1.I_B + \beta_2.\beta_1.I_B + \beta_2.I_B$$

$$I_C = \left(\beta_1 + (\beta_2.\beta_1) + \beta_2\right).I_B$$

Where β_1 and β_2 are the gains of the individual transistors.

This means that the overall current gain, β is given by the gain of the first transistor multiplied by the gain of the second transistor as the current gains of the two transistors multiply. In other words, a pair of bipolar transistors combined together to make a single Darlington transistor pair can be regarded as a single transistor with a very high value of β and consequently a high input resistance.

Darlington Transistor Example-1

A 12V 75W halogen bulb is switched using a Darlington pair made up of two NPN transistors. If the first transistor's forward current gain is 25 and the second's forward current gain (Beta) is 80. Calculate the maximum base current necessary to turn the light completely ON, taking into account any voltage drops across the two transistors. First, the lamp's current draw will be equal to the second transistor's Collector current:

$$I_C = I_{LAMP}$$

$$\therefore I_{LAMP} = \frac{P}{V} = \frac{75}{12} = 6.25\,Amps$$

Utilizing the equation above, the base current is given as:

$$\beta_1 = 25, \quad \beta_2 = 80$$

$$I_C = \left(\beta_1 + \left(\beta_2 . \beta_1\right) + \beta_2\right).I_B$$

$$\therefore I_B = \frac{I_C}{\beta_1 + \left(\beta_2 . \beta_1\right) + \beta_2} = \frac{6.25}{2105} = 3.0\,mA$$

It has been observed that a very small base current of only 3.0mA, such as that supplied by a digital logic gate or the output port of a micro-controller, can be used to switch the 75 Watt lamp "ON" and "OFF".

If two identical bipolar transistors are used to make a single Darlington device then β_1 is equal to β_2 and the overall current gain will be given as:

If $\beta_1 = \beta_2$

$$I_C = \left(\beta_1 + (\beta_2.\beta_1) + \beta_2\right).I_B$$

$$\therefore I_C = \left(\beta^2 + 2\beta\right).I_B$$

Generally the value of β^2 is much greater than that of 2β, in which case it can be ignored to simplify the maths a little. Then the final equation for two identical transistors configured as a Darlington pair can be written as:

Identical Darlington Transistors

$$I_C = \left(\beta^2 \times I_B\right)$$

It has been observed that for two identical transistors, β^2 is used instead of β acting like one big transistor with a huge amount of gain. Darlington transistor pairs with current gains of more than a thousand with maximum collector currents of several amperes are easily available. For example: the NPN TIP120 and its PNP equivalent the TIP125.

The benefit of using an arrangement such as this, is that the switching transistor is much more sensitive as only a tiny base current is required to switch a much larger load current as the typical gain of a Darlington configuration can be over 1,000 whereas normally a single transistor stage produces a gain of about 50 to 200.

It has been observed that a darlington pair with a gain of 1,000:1, could switch an output current of 1 ampere in the collector-emitter circuit with an input base current of just

1mA. This then makes darlington transistors ideal for interfacing with relays, lamps and motors to low power microcontroller, computer or logic controllers as shown.

Darlington Transistor Applications

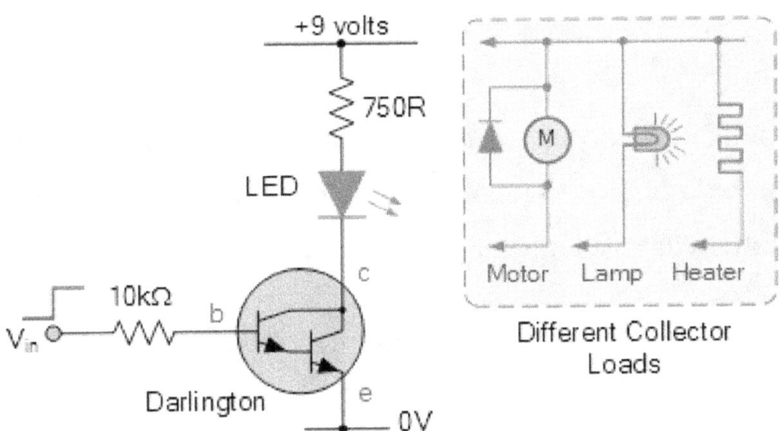

The base of the Darlington transistor is sufficiently sensitive to respond to any small input current from a switch or directly from a TTL or 5V CMOS logic gate. The maximum collector current Ic(max) for any Darlington pair is the same as that for the main switching transistor, TR_2 so can be used to operate relays, DC motors, solenoids and lamps, etc.

One of the main drawbacks of a Darlington transistor pair is the minimum voltage drop between the base and emitter when fully saturated. Unlike a single transistor which has a saturated voltage drop of between 0.3v and 0.7v when fully-ON, a Darlington device has twice the base-emitter voltage drop (1.2 V instead of 0.6 V) as the base-emitter voltage drop is the sum of the base-emitter diode drops of the two individual transistors which can be between 0.6v to 1.5v depending on the current through the transistor.

Due to the Darlington transistor's large base-emitter voltage drop, which makes it hotter than a typical bipolar transistor for a given load current, it needs effective heat dissipation. Darlington transistors also have slower ON-OFF response times because it takes longer for the slave transistor TR1 to fully ON or fully OFF the master transistor TR2.

A complimentary NPN and PNP transistor can be utilized in the same cascaded configuration to create a different sort of Darlington transistor known as a Sziklai Configuration to get around the delayed response, increased voltage drop, and thermal drawbacks of a regular Darlington Transistor device.

Sziklai Transistor Pair

The Sziklai Darlington Pair is a complementary or compound Darlington device made up of individual NPN and PNP complementary transistors coupled together as shown below. It is named after its Hungarian inventor George Sziklai.

This cascaded combination of NPN and PNP transistors has the advantage that the Sziklai pair performs the same basic function of a Darlington pair except that it only requires 0.6v for it to turn-ON and like the standard Darlington configuration, the current gain is equal to β^2 for equally matched transistors or is given by the product of the two current gains for unmatched individual transistors.

Configuration of Sziklai Darlington Transistor

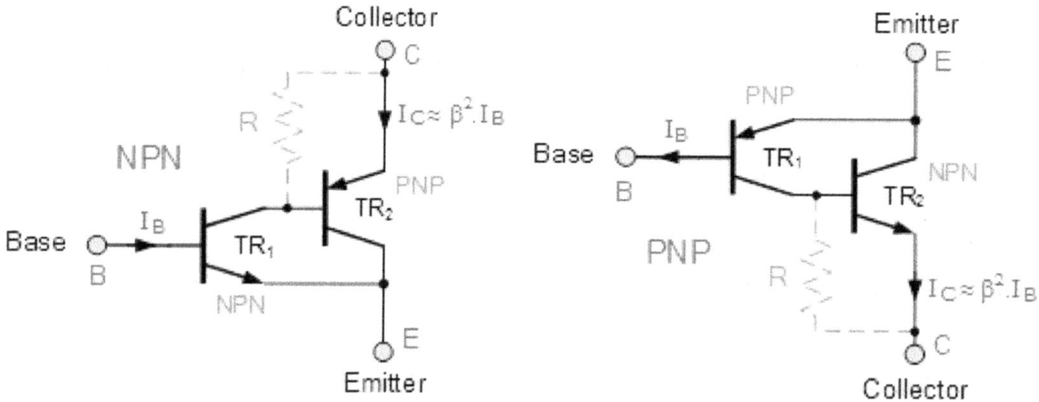

From the above circuit, it has been observed that the base-emitter voltage drop of the Sziklai device is equal to the diode drop of a single transistor in the signal path. However, the Sziklai configuration can not saturate to less than one whole diode drop, i.e. 0.7v instead of the usual 0.2v.

The Sziklai pair also has slower response times than a single transistor, just as the Darlington pair. Push-pull and class AB audio amplifier output stages frequently employ Sziklai pair complementary transistors, which only permit one polarity of the output transistor. There are NPN and PNP variations of the Darlington and Sziklai transistor pairs.

Darlington Transistor IC's

As some output devices, like LEDs or displays, only need a few milliamps to operate at low DC voltages and can therefore be driven directly by the output of a standard logic gate, it is sufficient for the controlling circuit to switch a DC output voltage or current "ON" or "OFF" in the majority of electronics applications. However, as we've seen above, a micro-controller or regular logic gate may not always be able to provide all the power needed to operate the output device, such as a DC motor.

In order to drive the device if the digital logic device is unable to deliver enough current, further circuitry will be needed. The ULN2003 array is one such widely used Darlington transistor device. The ULN2002A, ULN2003A, and ULN2004A, all high voltage, high current darlington arrays with seven open collector darlington pairs in a single IC package, make up the darlington array family.

The array's 500mA-rated channels each have a peak current limit of 600mA, making them perfect for controlling low-power semiconductor gates and bases as well as small motors or lamps. To make the connections and board layout simpler, more suppression diodes are added for inductive load driving, and the inputs are pinned opposite the outputs.

The ULN2003A Darlington Transistor Array

The ULN2003A is a low-cost unipolar darlington transistor array that can drive a variety of loads, such as solenoids, relays, DC motors, LED displays, or filament lamps, thanks to its excellent efficiency and low power consumption. Seven darlington transistor pairs, each having an input pin on the left and an output pin directly across from it on the right, are present in the ULN2003A.

ULN2003A Darlington Transistor Array

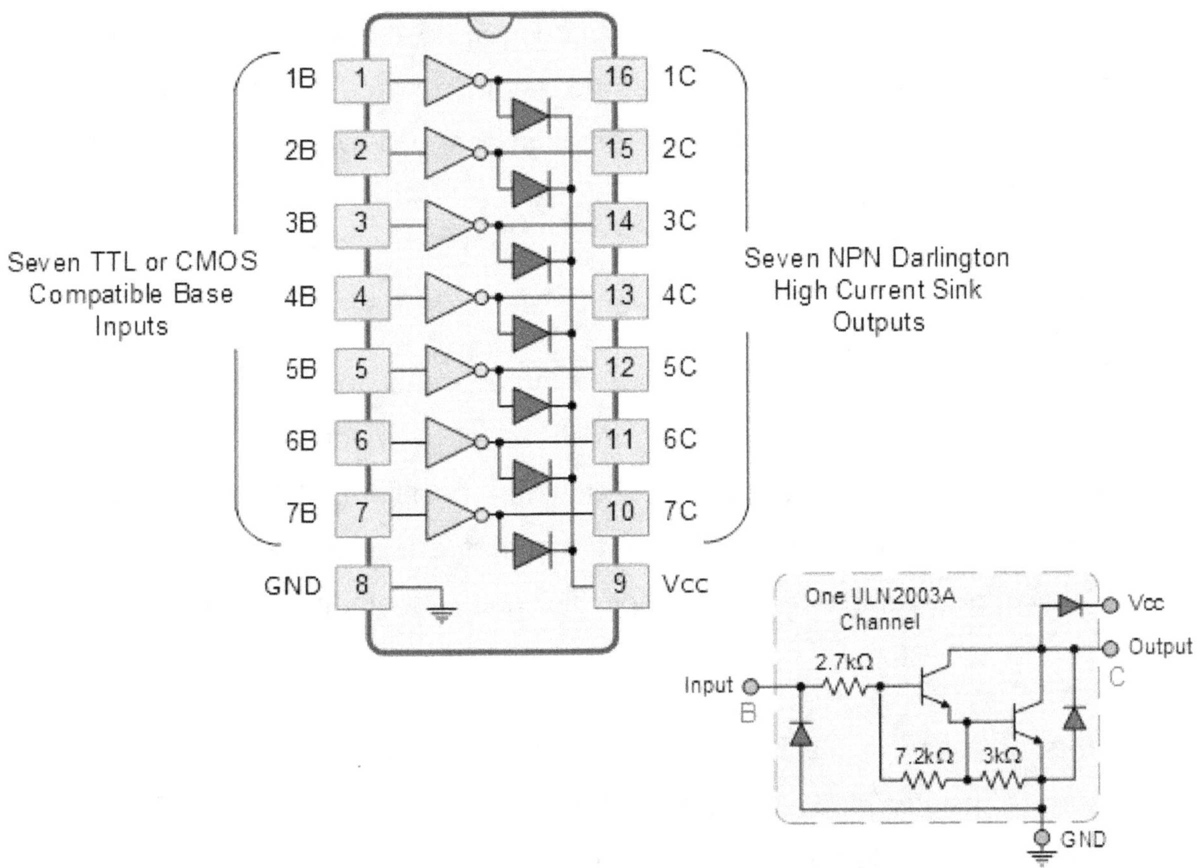

The ULN2003A Darlington driver may be powered directly from a TTL or +5V CMOS logic gate and has a very high input impedance and current gain. Use the ULN2004A for +15V CMOS logic, and the SN75468 Darlington array for larger switching voltages up to 100V. The associated output (pins 1 to 7) will switch to "LOW" sinking current when an input (pins 1 to 7) is driven "HIGH".

Similar to how the corresponding output changes to a high impedance state when the input is driven "LOW." The device's efficiency is increased by blocking load current and reducing leakage current in the high impedance "OFF" state.

While pin 9 (V_{cc}) is linked to the load's supply, pin 8 (GND) is connected to the load's ground, or 0 volts. Between $+V_{cc}$ and one of the output pins, pins 10 to 16, any load then needs to be attached. Pin 9 should always be connected to V_{cc} for inductive loads like as motors, relays, solenoids, etc.

The ULN2003A can switch 500mA (0.5A) per channel, however if additional switching current capability is needed, both the inputs and outputs of the Darlington pairs can be paralleled for greater current capability. As an illustration, to switch the load, connect input pins 1 and 2 and output pins 16 and 15.

Darlington Transistor

A high power semiconductor device called a Darlington transistor has individual current and voltage ratings that are far higher than those of a typical tiny signal junction transistor. In comparison to small signal switching transistors, the DC current gain values for standard high power NPN or PNP transistors are quite low, as low as 20 or even less. This implies that to switch a specific load, substantial base currents are needed.

Two transistors are used in the Darlington configuration, one of which is the primary current-carrying transistor and the other of which is a much smaller "switching" transistor that supplies the base current to drive the main transistor. Due to the fact that the DC current gains of the two transistors are multiplied together, a smaller base current can be utilized to switch a much higher load current.

The two-transistor combination can then be thought of as one single transistor with a very high value of β and a high input resistance as a result. Complementary Sziklai Darlington transistors, which combine separate matching NPN and PNP complementary transistors within a single Darlington pair to increase efficiency, are also available in addition to the normal PNP and NPN Darlington transistor pairs.

In robotic and mechatronic applications, Darlington arrays like the ULN2003A are also available, allowing high power or inductive loads like lights, solenoids, and motors to be securely operated by microprocessor and microcontroller devices.

Open Collector Outputs are excellent for switching incompatible loads, but to achieve the proper switching action, they may need a pull-up or pull-down resistor. Open Collector Outputs are becoming more and more common in applications involving operational amplifiers, micro-controllers (Arduino), and digital chip design. They can be used to interface with other circuits or to drive high-current loads like indicator lamps and relays that might not be compatible with the electrical properties of the control circuit.

How can we use "open-collector" in our circuit designs and what does it mean?

We know that an NPN or PNP type bipolar junction transistor is a 3-terminal device. The Emitter, Base, and Collector are the names given to these three terminals. Bipolar transistors can be used as an amplifier, in which case the output signal is stronger than the input signal, or, more frequently, as a solid state "ON/OFF" type electrical switch.

The Bipolar Junction Transistor (BJT), which has three terminals, can be set up and used in one of three switching modes. There are three of them: Common Base (CB), Common Emitter (CE), and Common Collector (CC). When utilized for switching or amplification (active region), the "Common Emitter" configuration is the most popular transistor operation (cut-off or saturation regions). Here, on open collector outputs, we will therefore focus on this transistor design. Think about the configuration of the common emitter amplifier that is displayed below.

Configuration of Common Emitter

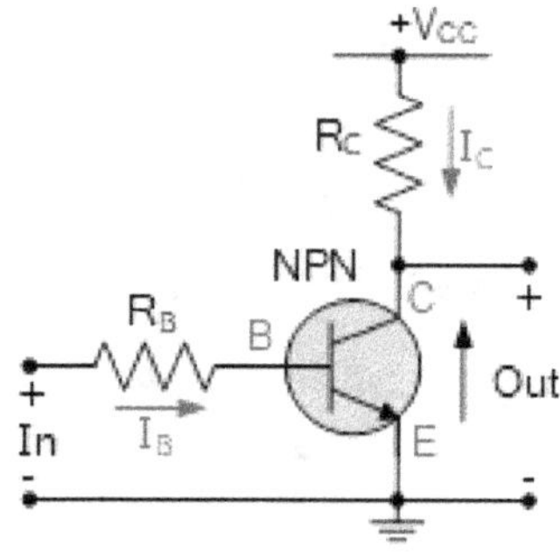

The transistor's collector terminal and the positive supply rail, V_{CC}, are connected by a resistance in this single stage common emitter configuration. The input signal is applied at the junction of the transistor's base and emitter, and the terminal of the emitter is directly linked to ground. Hence, "common emitter" as a descriptive phrase (CE).

The output signal, which is 180°-phase inverted compared to the input signal, is taken from between the collector and emitter terminals and fed directly into the base of the NPN transistor via base resistor R_B to turn the transistor "ON." This makes it possible to adjust the transistor's collector current between zero (cut-off) and a maximum value (saturation). The common emitter setup is typically set up in this way, either biased to function as a class-A amplifier or as a logical ON/OFF switch.

Since the transistor and its collector load resistance are connected to the same common supply voltage, this is an issue. The transistor can produce an amplified output signal by using the collector resistor, R_C, to enable the collector voltage, V_C, to change value in response to an input signal applied to the transistor's base terminal. As the voltage at the collector terminal would always equal the source voltage in the absence of R_C.

We know that, when V_{BE} is significantly less than 0.7 volts (zero base current) or significantly larger than 0.7 volts (maximum base current), a bipolar junction transistor can function between its cut-off and saturation areas. The NPN bipolar transistor can be used as an electronic switch that performs an inversion operation in this way because, when it is "OFF," its collector terminal, and therefore V_{CE}, are "HIGH" at V_{CC} level, and, when it is "ON," (conducting), the output taken across V_{CE} will be "LOW." These are the opposite switching conditions if we want to control a relay, solenoid, or lamp, for example.

Removing the collector resistor, R_C, entirely and leaving the transistor's collector terminal open to be linked to an external load is one technique to get around this inversion of the switching state. An open collector output configuration is what is produced by this kind of design.

NPN Open Collector Output

It can be totally ON or entirely OFF when an NPN bipolar transistor is operated in an Open Collector (OC or o/c) mode, operating as an electrical solid-state switch. In other words, the transistor will be entirely OFF when no base bias voltage is applied, and fully ON when an appropriate base bias voltage is applied.

The transistor does not function as an amplifier as it would if it were regulated in its active zone when it is switched between its cut-off (OFF) and saturation regions (ON). The ability of the open collector outputs to drive externally connected loads requiring higher voltages and/or currents than those permitted by the prior common emitter arrangement is made possible by the transistor's capacity to transition between cut-off and saturation. The actual switching transistor's maximum permitted voltage and/or current values serve as the only restriction.

The benefit of open collector outputs then is that any switching voltage for the output can be produced by simply connecting the collector terminal to the single positive supply as before or by powering the load from a different supply rail. For instance, you could want to use the output pin of an Arduino, Raspberry-Pi, or +5 volt logic gate to power a low-current bulb or relay that needs a +12 volt source.

The drawback is that since the transistor's collector terminal lacks the capacity for output drive, an external pull-up resistor is typically needed when using open collector outputs to switch digital signals, gates, or inputs of electronic circuits. This is due to the fact that an NPN transistor can only draw the output LOW to ground (0V) when powered up; when it is turned off, it cannot push the output HIGH once more.

In order to prevent the open collector terminal from oscillating between HIGH (+V) and LOW (0V) when the transistor is OFF, the output must be pulled HIGH again when the transistor is de-energized using an external "pull-up resistor" linked between its collector terminal and the supply voltage. The pull-up resistor's value isn't crucial and will vary slightly depending on the load current required at the output; common resistive values range from a few hundred to a few thousand ohms. As a result, the main purpose of an NPN bipolar transistor's open-collector outputs is to sink current.

Open Collector Transistor Circuit

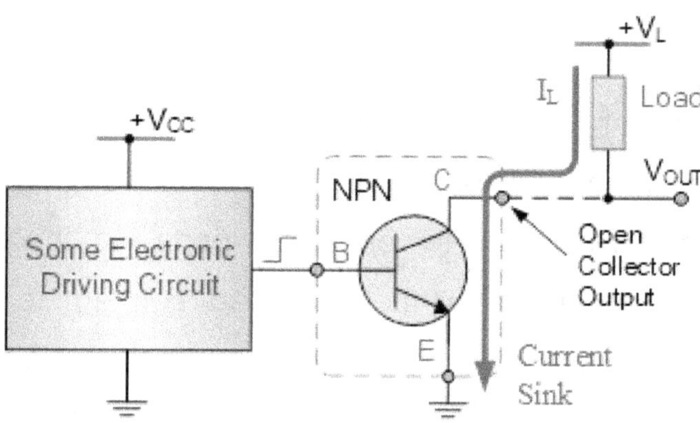

An open collector switching circuit, which is helpful for controlling electromechanical devices as well as many other switching applications, is shown in the image above in its typical configuration. Any suitable analog or digital circuit could be used to drive the base of an NPN transistor. The load that will be switched is connected to the transistor's collector, while ground is linked to the transistor's emitter terminal.

When a control signal is applied to the base of a transistor with an NPN-type open collector output, the transistor turns ON. The output, which is connected to the collector terminal, is then pulled down to ground potential through the now-conducting transistor junctions, energizing and turning ON the connected load. Thus, using Ohm's Law to compute the load current, I_L, the transistor switches and passes it as follows:

Load Current, I_{load} = Voltage across load / Resistance of load

The load, which could be a relay coil, solenoid, small dc motor, lamp, etc., is de-energized and also goes OFF when the transistor's positive base drive is removed (OFF). Once the current-sink switching action of the NPN transistors open-collector functions as either an open circuit (OFF) or a short circuit, the output transistor can be utilized to control an externally connected load (ON).

The benefit of this approach is that the collector load need not be linked to the same voltage potential as the transistor driving circuit; instead, it can operate at a lower or higher voltage potential, such as 12 volts or 30 volts DC. Additionally, by just swapping out the output transistor, the same straightforward digital or analog circuit can be used to switch a variety of loads. Use an open collector Darlington transistor, for instance, or a 2N3904 transistor operating at 6 VDC and 10 mA or a 2N3506 transistor operating at 40 VDC and 3 amperes.

Example-1 of Open Collector Output

As part of a school project, an electromechanical relay must be driven by a +5 volt digital output pin from an Arduino board. Calculate the base resistor needed to run the relay coil if its coil is rated at 12 VDC, 100, and the NPN transistor being utilized has a DC current gain (Beta) value of 50.

The current through the coil can be calculated using Ohm's law as: $I = V/R$

$$I_C = \frac{V_L}{R_L} = \frac{12}{100} = 0.12\,A \ \text{ or } \ 120\,mA$$

$$\beta = \frac{I_C}{I_B} \quad \therefore I_B = \frac{I_C}{\beta}$$

$$I_B = \frac{I_C}{\beta} = \frac{0.12}{50} = 2.4\,mA$$

Therefore, ignoring the collector-emitter saturation voltage ($V_{CE(sat)}$) of approximately 0.2 volts, a base current of 2.4 mA is needed for an NPN transistor with a DC current gain of 50. Remember that the DC current gain of a transistor specifies the amount of base current needed to generate the consequent collector current. The transistor will have a 0.7 volt voltage drop across the base-emitter junction (V_{BE}) when it is completely ON.

Therefore, the needed base resistor's value, R_B, is determined as follows:

$$R_B = \frac{V_{RB}}{I_B} = \frac{(5v - 0.7v)}{0.0024} = 1792\,\Omega \ \text{ or } \ 1.8k\Omega$$

Therefore, the open-collector transistor circuit would be:

Open Collector Circuit

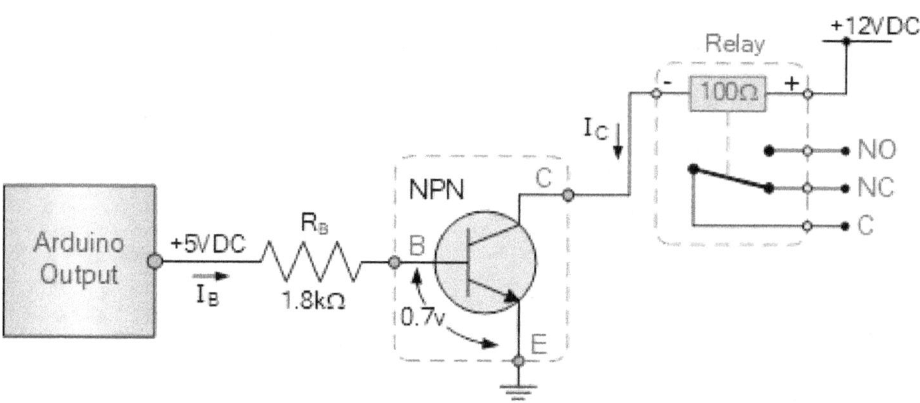

A PNP-type transistor can also be used in an open collector configuration to produce what is known as a "current-sourcing" output, whereas an NPN open collector transistor circuit produces a "current-sinking" output, meaning that the NPN transistor's open collector terminal will sink the current to ground (0V).

PNP Open Collector Output

As we saw above, an open-collector output's primary characteristic is that the load signal is actively "drawn down" to ground level by the switching action of the NPN bipolar transistor when it is fully ON and passively pushed back up again when it is fully OFF, producing a current sink output.

However, by utilizing a PNP bipolar transistor's open collector output to actively switch its output toward a voltage supply rail and an externally connected "pull-down" resistor to passively draw the output down again when OFF, we may produce the reverse switching situation.

Since the transistor can only switch the output HIGH to the supply rail for a PNP-type open collector output, its output terminal must be passively pushed "LOW" once again by an externally attached "pull-down" resistor, as shown.

Open Collector PNP Transistor Circuit

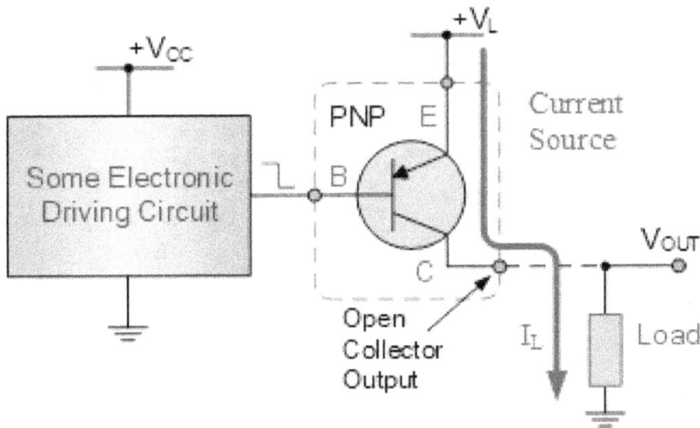

Therefore, it becomes clear that an NPN-type or PNP-type open collector output configuration can only actively pull its output LOW to ground or HIGH to a supply rail (depending on the transistor type) when ON, and that if the connected load is unable to do so, the collector terminal must be passively pulled up or down by means of a pull-up or pull-down resistor connected to its output terminal.

The switching activity of the output transistor determines whether a current sink or source state is produced. It is feasible to employ n-channel and p-channel enhancement mode MOSFETs or IGBTs in their open-source configuration in addition to bipolar transistors in their open collector configuration. The normally-open (enhancement) MOSFET requires an appropriate voltage given to its gate (G) terminal, in contrast to the bipolar junction transistor (BJT), which requires a base current to force the transistor into saturation.

The open-drain (D) terminal of the MOSFET is connected to the external load, while the source (S) terminal is directly connected to ground or the supply rail. When driving power loads or loads connected to a higher voltage source, MOSFETs (or IGBTs) used as open-drain (OD) devices must adhere to the same specifications as open-collector outputs (OC), in that pull-up or pull-down resistors must be used. The MOSFETs' channel thermal power rating and static voltage protection are the only things that change.

Configuration of Open Drain Enhancement MOSFET

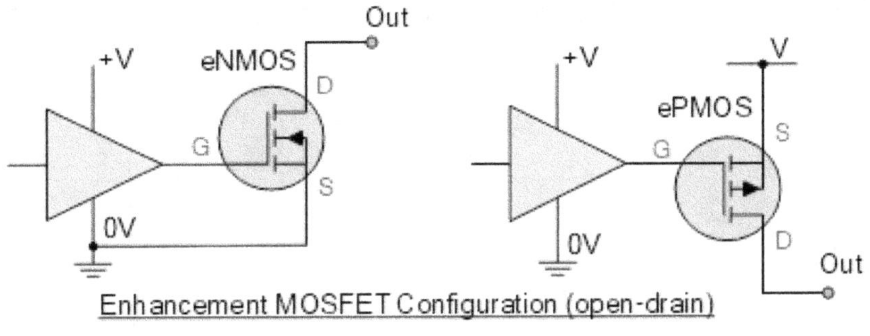

<u>Enhancement MOSFET Configuration (open-drain)</u>

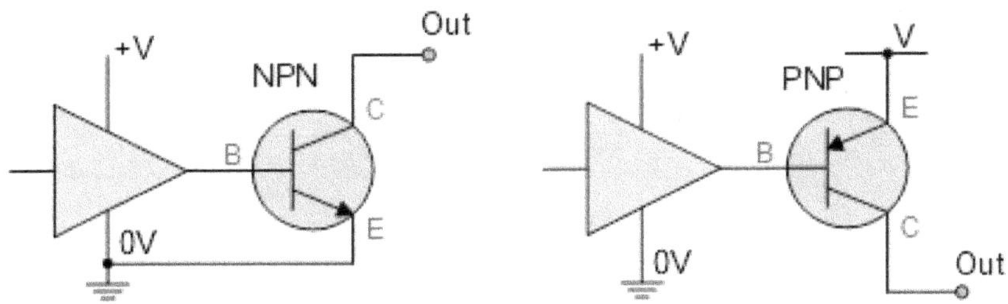

<u>Bipolar Transistor Configuration (open-collector)</u>

Open Collector Output in A Brief

Depending on the type of bipolar transistor used—NPN-type or PNP-type—the open collector output can provide either a current sink output or a current source output. An NPN-type transistor will create a channel to ground while it is in the "ON" state, or "sink."

If the open collector output isn't linked to the positive supply voltage using a pull-up resistor, the output terminal may float when the device is in the "OFF" state. A PNP-type transistor operates in the contrary manner. It will supply or "source" a path from the supply rail when it is in its "ON" state. Unless the open-collector output is connected to ground using a pull-down resistor when the device is in the "OFF" state, its output terminal may float (0V).

Open collector outputs and open drain outputs have the benefit of allowing the load to be switched or controlled to be connected to a voltage supply that is independent of and/or different from the supply voltage used by the controlling circuit. They also have the advantage of being able to "sink" or "source" an externally supplied voltage depending on whether they are connected to ground or a source. The output switching transistor's or e-maximum MOSFET's permissible voltage and current ratings are the only restriction.

CHAPTER-10: FET CURRENT SOURCE

JFETs and MOSFETs are used in FET constant current sources to deliver a load current that stays constant despite variations in supply voltage or load resistance. An active circuit known as a FET Current Source uses a Field Effect Transistor to provide a steady stream of current to a circuit.

But why would you choose a steady stream of current?

It is quite easy to create biasing circuits or voltage references with constant current values, such as 100uA, 1mA, or 20mA with only a single FET and resistor, by utilizing constant current sources and current sinks (a current sink is the opposite of a current source).

In capacitor charging circuits for precise timing purposes or in rechargeable battery charging applications, as well as in linear LED circuits for driving strings of LEDs at a constant brightness, constant current sources are frequently utilized.

Constant current sources can also be used to provide resistive voltage references because, if the resistance value is known and the current flowing through it is steady and constant, the voltage drop can be calculated using Ohm's law.

However, employing low transconductance FETs and precise resistor values to transform the current into a precise and steady voltage is essential for building an accurate and dependable FET-based constant current source.

In addition to Junction-FETs (JFETs) and Metal-oxide Semiconductor MOSFETs, which are already employed in low current-source applications, field-effect transistors are frequently used to provide a current source. The JFET can be used as a voltage-controlled resistor in its most basic configuration, with a modest gate voltage controlling the channel's conduction.

Biasing the Junction FET

The N-channel JFET is a "normally-ON" device until the gate-to-source voltage (VGS) gets negative enough to switch it "OFF". JFETs are depletion devices. The P-channel JFET, a depletion device that is also "usually ON," needs the gate voltage to reach sufficiently positive to make it "OFF."

N-channel JFET Biasing

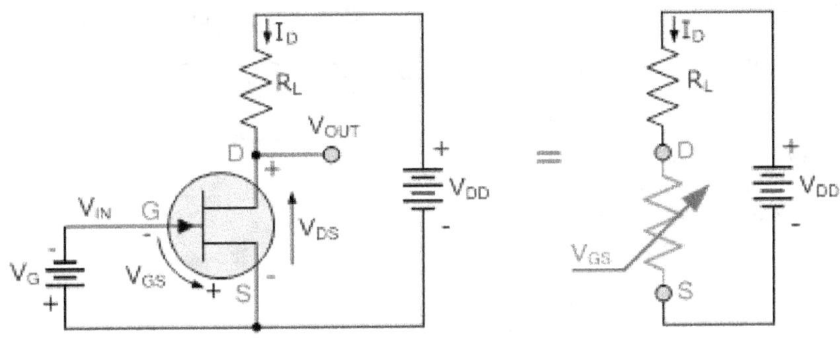

The illustration displays the typical setup and connections for an N-channel JFET with a common source configuration and normal biasing when utilized in its active area. In this case, the gate-source voltage V_{GS} is equal to the gate supply, or input voltage V_G, which establishes the reverse bias between the gate and the source. Meanwhile, V_{DD} supplies the voltage from the drain to the source, allowing current to flow from the source to the drain. This current carrying an I_D label enters the JFET drain terminal.

For various gate-source values of V_{GS}, the forward voltage drop of the JFET is represented by the drain-source voltage V_{DS}, which depends on the drain current I_D. The drain-to-source saturation current $I_{D(sat)}$, also known as the I_{DSS}, is the current that flows when V_{DS} is at its minimum value, the JFET's conductive channel is completely open, and I_D is at its greatest value.

The JFET's conductive channel is completely closed (pinched off) when V_{DS} reaches its maximum value, which causes I_D to decrease to zero with the drain-to-source voltage, V_{DS}, which is equal to the drain supply voltage V_{DD}. The gate cut-off voltage V_{GS} is the gate voltage at which the JFET's channel ceases to conduct (off).

Since V_{GS} and I_D are steady-state quantities, the N-channel JFET's common source biasing configuration controls how the JFET will operate in the quiescent state, or steady state, when there is no input signal. As a result, a common-source JFET is a voltage-controlled device since its input voltage affects the amount of current that flows through its conductive channel between its drain and source (V_{GS}). Therefore, by graphing I_D vs V_{GS} for every certain JFET device, we may create a set of output characteristic curves.

N-channel JFET Output Characteristic

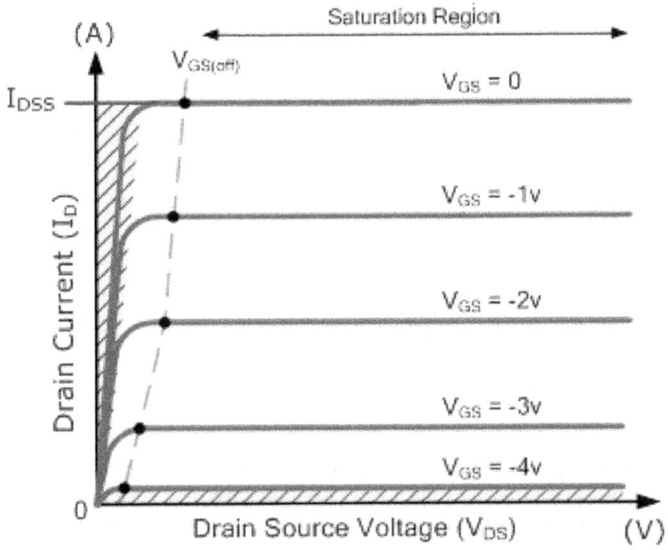

The JFET as a Constant Current Source

The n-channel JFET is a normally-ON device, therefore we might use this since, if V_{GS} is sufficiently negative, the drain-source conductive channel closes (cuts off), reducing the drain current to zero. The p-type depletion region surrounding the gate for an n-channel JFET widens until it entirely blocks the channel, which causes the conductive channel between the drain and source to collapse.

For a p-channel JFET, N-type depletion patches close the channel. Therefore, we can make the JFET a perfect FET current source by causing it to conduct current through its channel at a certain value between zero amperes and I_{DSS}, respectively, by setting the gate-source voltage to some pre-determined fixed negative value. Take a look at the circuit below.

JFET Zero-voltage Biasing

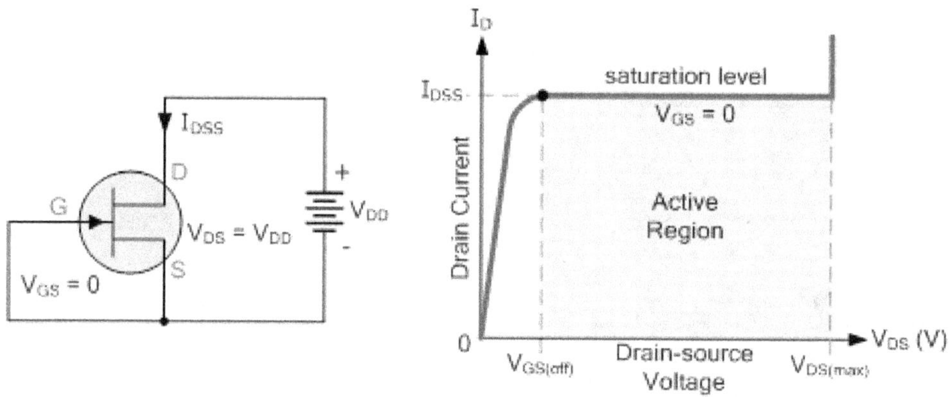

We observed that the output characteristic curves of the JFET are a plot of I_D vs V_{GS} for a fixed V_{DS}. However, we also observed that even with significant variations in V_{DS}, the JFET curves hardly alter, and this parameter can be very helpful in determining a constant operating point for the conductive channel.

The conductive channel of the JFET is open and the gate terminal is shorted to the source terminal, creating the simplest constant FET current source. Since the JFET is operating in its saturated current area, the current flowing through it will be close to its maximum I_{DSS} value.

However, since the JFET is always in full conduction and the I_{DSS} current value varies greatly depending on the device type, the operation and performance of such a constant current setup is really subpar. The n-channel J1xx or PN4xxx series, for instance, can be several tens of milli-amperes, whereas the smaller 2N36xx or 2N43xx series is only a few mill-amperes (mA).

Also keep in mind that, despite what manufacturers claim on their data sheets, the lowest and maximum values of this zero gate voltage drain current, I_{DSS}, will vary significantly between devices with the same part number.

Another thing to keep in mind is that a FET is essentially a voltage-controlled resistor with a resistive value placed in series with the drain and source terminals on its conductive channel. R_{DS} is the name of this channel resistance. As we have seen, the maximum drain-to-source current flows when $V_{GS} = 0$, hence the channel resistance of the JFET, R_{DS}, must be at its lowest.

Although it can be as high as 50 or so Ohms, the channel resistance is not absolutely zero but rather has a low ohmic value determined by the FET's manufacturing shape. This channel resistance, sometimes referred to as $R_{DS(ON)}$ when a FET is conducting, is at its lowest resistive value when $V_{GS} = 0$. Thus, a low I_{DSS} and a high $R_{DS(ON)}$ number are equivalent.

Therefore, when V_{GS} is equal to zero volts, a JFET can be biased to act as a FET current source device at any current value below its saturation current, I_{DSS}. There will be no drain current ($I_D = 0$) since the channel will be closed when V_{GS} reaches its $V_{GS(off)}$ cut-off voltage level. Therefore, as long as the JFET device is operated within its active zone, as indicated, the channels drain current, I_D, will always flow.

JFET Transfer Curve

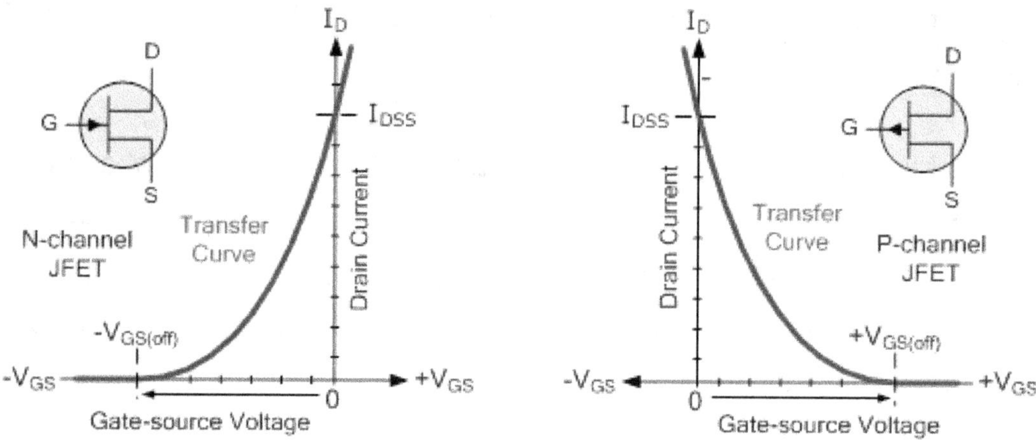

It should ne noted here that for a P-channel JFET, the $V_{GS(off)}$ cut-off voltage will be a positive voltage but its saturation current, I_{DSS} obtained when V_{GS} equals zero volts will be the same as for an N-channel device. Also notice that the transfer curve is nonlinear because the drain current is increasing faster through the opening channel as V_{GS} approaches zero volts.

JFET Negative-voltage Biasing

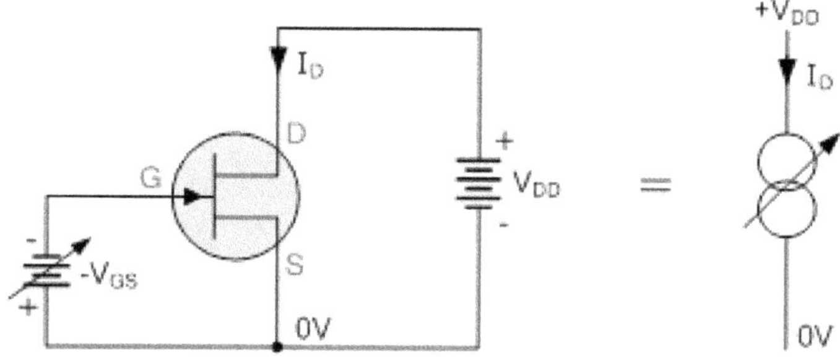

As a depletion mode device, the JFET is always "ON," hence to turn it "OFF," an N-channel JFET needs a negative gate voltage, and a P-channel JFET needs a positive gate voltage. An N-channel JFET biased positively or a P-channel JFET biased negatively will further open the conductive channel and force the channel current, I_D, to exceed I_{DSS}. However, if we use the characteristic curves of I_D against V_{GS}, we may change V_{GS} to a negative voltage level, such as -1V, -2V, or -3V, to provide a fixed JFET constant current source that can produce any level of current between zero and I_{DSS}.

But it is preferable to bias the JFET at roughly 10% to 50% of its maximum I_{DSS} value for a constant current source that is more precise and has better regulation. Additionally, by reducing I^2*R power losses through the resistive channel, the effect of heating is reduced.

We can see that a JFET's operating point, which permits the channel to conduct and pass a specific value of drain current, I_D, can be established by biasing the gate terminal with some negative voltage value, or a positive voltage for a P-channel JFET. A JFET drain current I_D can be mathematically written as: for various values of V_{GS}.

JFET Drain Current Equation

$$I_D \cong I_{DSS}\left[1 - \frac{V_{GS}}{V_{GS(off)}}\right]^2$$

Example-1 of FET Constant Current Source

The manufacturers datasheet for a J107 N-channel switching JFET shows that it has an I_{DSS} of 40mA when $V_{GS} = 0$, and a maximum $V_{GS(off)}$ value of -6.0 volts. Using these declared values, calculate the JFET's drain current value when, $V_{GS} = 0$, $V_{GS} = -2$ volts, and when $V_{GS} = -5$ volts. Also show the J107's transfer characteristic curve.

1). When $V_{GS} = 0V$

When $V_{GS} = 0V$ the conductive channel is open and maximum drain current flows.

Thus $I_D = I_{DSS} = 40mA$.

2). When V_{GS} = -2V

$$I_D \cong I_{DSS}\left[1 - \frac{V_{GS}}{V_{GS(off)}}\right]^2 = 40\,\text{mA}\left[1 - \frac{-2}{-6}\right]^2$$

$$I_D = 0.04(1-0.333)^2 = 0.04(0.444) = 17.78\,\text{mA}$$

3). When V_{GS} = -5V

$$I_D \cong I_{DSS}\left[1 - \frac{V_{GS}}{V_{GS(off)}}\right]^2 = 40\,\text{mA}\left[1 - \frac{-5}{-6}\right]^2$$

$$I_D = 0.04(1-0.833)^2 = 0.04(0.0278) = 1.11\,\text{mA}$$

4). J107 Transfer Characteristic Curve

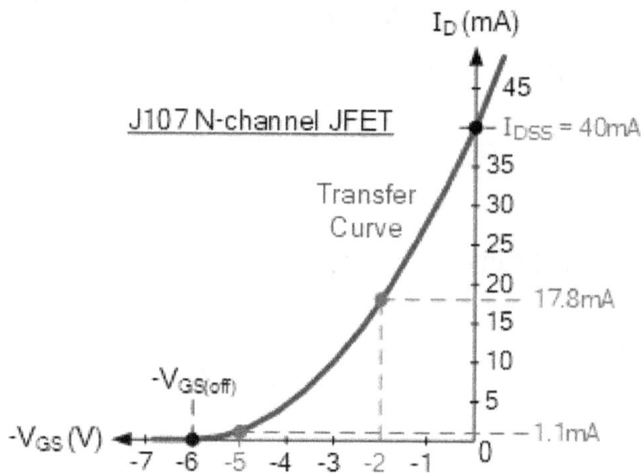

Therefore, it has been observed that as the gate-source voltage, V_{GS} approaches the gate-source cut-off voltage, $V_{GS(off)}$ the drain current, I_D decreases. In this simple example, we calculated the drain current at two points, but calculating using additional values of V_{GS} between zero and cut-off would give us a more accurate shape of the curve.

JFET Current Source

A JFET can be made to operate as a voltage controlled constant current source whenever its gate-source junction is reverse biased, and for an N-channel device we need a $-V_{GS}$ and for a P-channel device we need a $+V_{GS}$. The problem here is that the JFET requires two separate voltage supplies, one for V_{DD} and another for V_{GS}.

However, if we place a resistor between the source and ground (0 volts), we can achieve the necessary V_{GS} self-biasing arrangement for the JFET to operate as a constant current source using only the V_{DD} supply voltage. Consider the circuit below.

JFET Current Source

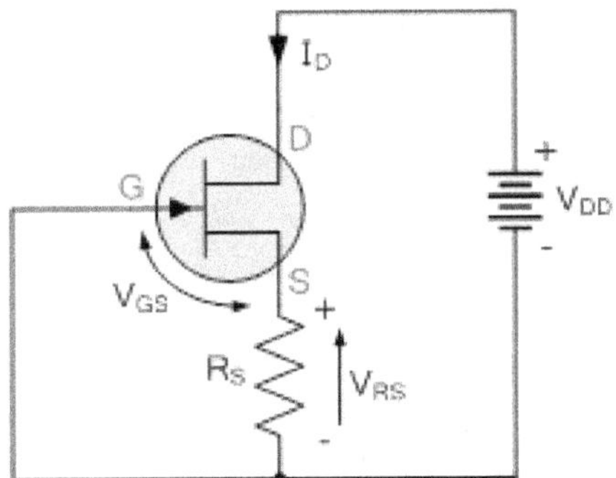

At first glance you may think that this configuration looks very similar to a JFET common drain (source follower) circuit. However, the difference this time is that while the FET's gate terminal is still tied directly to ground (V_G = 0), the source terminal is at some voltage level above zero voltage ground due to the voltage drop across the source resistor, R_S.

Therefore, with a channel current flowing through the external source resistor, the gate-to-source voltage of the JFET will be less than (more negative than) zero ($V_{GS} < 0$).

The external source resistor, R_S provides a feedback voltage which is used to self-bias the JFET's gate terminal keeping the drain current constant through the channel despite any changes in the drain-source voltage. Hence, the only voltage source we need is the supply voltage V_{DD} to provide the drain current and bias.

Therefore, the JFET uses the voltage drop across source resistor (V_{RS}) to set the gate bias voltage V_{GS} and therefore the channel current as we have seen above. Thus, increasing the resistive value of R_S will decrease the channels drain current I_D, and vice versa. But if we wanted to construct a JFET constant current source circuit, what would be a suitable value for this external source resistor, R_S.

Manufacturers data sheets for a particular N-channel JFET will give us the values of $V_{GS(off)}$ and I_{DSS}. Knowing the values of these to two parameters we can transpose the above JFET equation for the drain current, I_D to find the value of V_{GS} for any given value of drain current, I_D between zero and I_{DSS} as shown.

JFET Gate to Source Voltage Equation

$$I_D = I_{DSS}\left[1 - \frac{-V_{GS}}{-V_{GS(off)}}\right]^2$$

$$\therefore V_{GS} = -V_{GS(off)}\left[1 - \sqrt{\frac{I_D}{I_{DSS}}}\right]$$

Having found the gate-to-source voltage required for a given drain current, the value of the source biasing resistor value required is found by simply using Ohm's law, as R = V/I. Therefore:

JFET Source Resistor Equation

$$R_S = \frac{V_{GS}}{I_D} \ (\Omega\text{'s})$$

Example-2 of FET Constant Current Source

Utilizing the J107 N-channel JFET device from above which has an I_{DSS} of 40mA when $V_{GS} = 0$, and a maximum $V_{GS(off)}$ value of -6.0 volts. Determine the value of the external source resistor required to produce a constant channel current of 20mA and again for constant current of 5mA.

1). V_{GS} for I_D = 20mA

$$V_{GS} = -V_{GS(off)} \left[1 - \sqrt{\frac{I_D}{I_{DSS}}} \right]$$

$$V_{GS} = -6 \left[1 - \sqrt{\frac{20\,\text{mA}}{40\,\text{mA}}} \right] = -6(1 - 0.7071)$$

$$\therefore V_{GS} = -6 \times 0.2929 = -1.75 \text{ volts}$$

$$R_{DS} = \frac{V_{GS}}{I_D} = \frac{1.75\text{V}}{20\text{mA}} = \frac{1.75}{0.02} = 87.5\,\Omega$$

2). V_{GS} for $I_D = 5mA$

$$V_{GS} = -V_{GS(off)}\left[1 - \sqrt{\frac{I_D}{I_{DSS}}}\right]$$

$$V_{GS} = -6\left[1 - \sqrt{\frac{5mA}{40mA}}\right] = -6(1 - 0.3536)$$

$$\therefore V_{GS} = -6 \times 0.6464 = -3.88 \text{ volts}$$

$$R_{DS} = \frac{V_{GS}}{I_D} = \frac{3.88V}{5mA} = \frac{3.88}{0.005} = 776\,\Omega$$

Therefore, when $V_{GS(off)}$ and I_{DSS} are both known, we can use the above equations to find the source resistance required to bias the gate voltage for a particular drain current, and in our simple example this was 87.5Ω at 20mA, and 776Ω at 5mA. Hence, the addition of an external source resistor allows for the adjustment of the current source output.

If we replace the fixed value resistors with a potentiometer, we can make the JFET constant current source fully adjustable. For example, we could replace the two source resistors in the above example with one 1kΩ potentiometer, or trimmer. Correspondingly, as well as being fully adjustable, this JFET constant current source circuits drain current will remain constant even with changes in V_{DS}.

Example-3 of FET Constant Current Source

An N-channel JFET is needed to vary the brightness of a 5mm round red LED load between 8mA and 15mA. If the JFET constant current source circuit is fed from a 15 volt DC supply, determine the JFET's source resistance required to illuminated the LED between minimum and maximum brightness when the switching JFET has a maximum $V_{GS(off)}$ value of -4.0 volts and an I_{DSS} of 20mA when V_{GS} = 0. Draw the circuit diagram.

1). V_{GS} for I_D = 8mA

$$V_{GS} = -V_{GS(off)}\left[1 - \sqrt{\frac{I_D}{I_{DSS}}}\right]$$

$$V_{GS} = -4\left[1 - \sqrt{\frac{8mA}{20mA}}\right] = -4(1 - 0.6325)$$

$$V_{GS}(8mA) = -4 \times 0.3675 = -1.47 \text{ volts}$$

$$\therefore R_{DS(8mA)} = \frac{V_{GS}}{I_D} = \frac{1.47V}{8mA} = \frac{1.47}{0.008} = 184\Omega$$

2). V_{GS} for $I_D = 15mA$

$$V_{GS} = -V_{GS(off)}\left[1 - \sqrt{\frac{I_D}{I_{DSS}}}\right]$$

$$V_{GS} = -4\left[1 - \sqrt{\frac{15mA}{20mA}}\right] = -4(1 - 0.866)$$

$$V_{GS}(15mA) = -4 \times 0.134 = -0.536 \, volts$$

$$\therefore R_{DS(15mA)} = \frac{V_{GS}}{I_D} = \frac{0.536V}{15mA} = \frac{0.536}{0.015} = 36\Omega$$

Therefore, an external potentiometer is required which is capable of varying its resistance between 36Ω and 184Ω. The nearest preferred potentiometer value would be 200Ω.

Adjustable JFET Constant Current Source

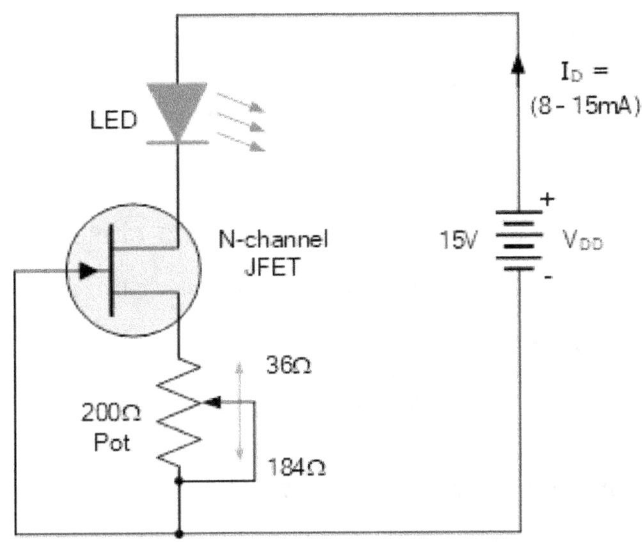

Instead of using a source resistance with a set value, R_S would allow us to adjust or adjust the current passing through the conductive channel of a JFET. However, it would be preferable to keep the maximum channel current flowing through the LED (15 mA in this case) between 10% and 50% of the JFETs I_{DSS} value in order to guarantee good current regulation through the FET device and a steadier current flow.

As opposed to JFETs, which are only available as normally-on depletion mode devices, MOSFETs are available in both depletion-mode (normally-on) and enhancement-mode (normally-off) devices as either P-channel or N-channel types, allowing for a greater range of current source options. Creating constant current sources using MOSFETs also allows for much greater channel currents and better current regulation.

FET Constant Current Source in A Brief

Field effect transistors can be used to supply a constant current to a load due to their channel resistance characteristics, as we have observed here bout the FET Constant Current Source, and they have many uses in electronic circuits where it is necessary to supply a fixed current to a connected load.

Depletion mode FETs, BJTs (bipolar junction transistors), or a mix of these two components can be used to build constant current circuits. Keeping in mind that, unlike the bipolar junction transistor, the JFET operates under voltage control. The fact that a

Junction Field Effect Transistor, or JFET, is a discharge device and that it needs a gate-to-source voltage, or V_{GS}, to switch it "OFF," is one of its key properties.

An N-channel JFET needs a $V_{GS(off)}$ voltage between 0 volts for full channel conduction and a negative value, typically several volts, for the JFET to be fully-OFF and close the channel. Therefore, we may regulate the channel depletion layers width and, consequently, its resistive value, passing a fixed and constant amount of current, by biasing the JFET's gate terminal at some fixed value between zero and $V_{GS(off)}$.

A P-channel JFET's $V_{GS(off)}$ value might be anywhere from zero volts for full channel conduction to multiple positive volts for a certain V_{DS} value. The amount of drain current, I_D that flows through the channel determines the regulation and tolerance of the constant current for a specific JFET device. The better the regulation, the lower the drain current via a particular device. The regulation and performance of a JFET can be enhanced by biasing it between 10% and 50% of its maximum I_{DSS} value.

An external resistance is connected between the source and gate terminals to accomplish this. The JFET can function as a continuous current source at any current level considerably below its saturation current, I_{DSS}, thanks to a gate-to-source feedback resistor, as shown above. This external source resistance, R_S, may be fixed or variable depending on the potentiometer used.

The key ideas in this transistors lesson segment can be summed up as follows:

After examining the design and functioning of field effect transistors (FETs), both junction and insulated gate, as well as NPN and PNP bipolar junction transistors (BJTs), we may summarize the key points of the transistor as follows:

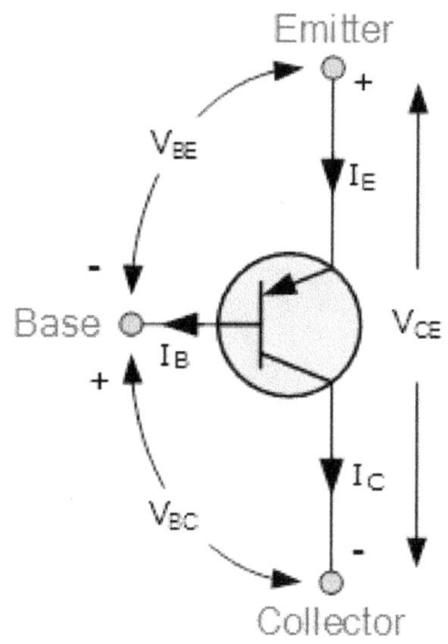

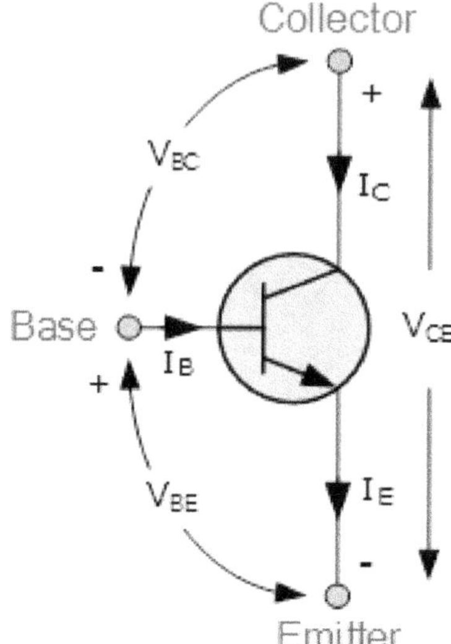

- The Bipolar Junction Transistor (BJT) is a three-layer electronic component made of two junctions of semiconductor diodes, one of which is forward biased and the other is reverse biased.
- NPN and PNP transistors are the two primary varieties of bipolar junction transistors (BJT).
- Bipolar junction transistors are "Current Operated Devices" in which a far greater Emitter to Collector current, which is itself almost equal, flows as a result of a much smaller Base current.
- The conventional current flow is depicted by the arrow in the transistor symbol.
- Common Emitter (CE) configuration is the most used transistor connection, but Common Base (CB) and Common Collector (CC) configurations are also possible.
- Requires a biasing voltage to operate an AC amplifier.
- The Base-Emitter junction is always forward biased whereas the Collector-Base junction is always reverse biased.
- The standard equation for currents flowing in a transistor is given as: $I_E = I_B + I_C$
- The Collector or output characteristics curves can be used to find either Ib, Ic or β to which a load line can be constructed to determine a suitable operating point, Q with variations in base current determining the operating range.

- Inductive loads such as DC motors, relays, and solenoids require a reverse biased "Flywheel" diode positioned across the load.
- A transistor can also be used as an electronic switch to control devices such as lamps, motors, and solenoids, etc. between its saturation and cut-off regions. This helps shield the transistor from any induced back emfs that may be produced when the load is turned "OFF."
- While the PNP type demands that the Emitter be more positive than the Base, the NPN type requires that the Base be more positive than the Emitter.

The Field Effect Transistor

- Field Effect Transistors, also known as FETs, are "Voltage Operated Devices" that fall into one of two categories:
- Devices with junction-gates are referred to as JFETs, and those with insulated-gates are referred to as IGFETs or, more often, MOSFETs.
- There are two other classifications for insulated-gate devices: enhancement kinds and depletion types.
- There are N-channel and P-channel variations of every form.
- FETs are excellent for use as electronic switches because they have very high input resistances, which result in very little or no current (MOSFET kinds) flowing into the input terminal.
- Since of the insulating oxide layer, the input impedance of the MOSFET is much higher than that of the JFET. Because of this, handling MOSFET devices requires caution because static electricity can quickly destroy them. An enhanced FET's transistor is in the "OFF" state, which is the same as an "open switch," when no voltage is supplied to the gate.
- Like a "closed switch," the depletion FET is naturally conductive and in the "ON" state when no voltage is provided to the gate.
- FETs outperform bipolar junction transistors in terms of current gains.
- The Common Source (CS) arrangement is the most typical FET connection; however Common Gate (CG) and Common Drain (CD) versions are also offered.

- Due to their extremely high channel "OFF" resistance and low "ON" resistance, MOSFETS make excellent switches.
- The gate of the N-channel JFET transistor needs to be applied a negative voltage in order to turn it "OFF".
- Positive voltage applied to the gate will "OFF" the P-channel JFET transistor.
- When a negative voltage is given to the gate to form the depletion area, N-channel depletion MOSFETs are in the "OFF" state.
- When a positive voltage is provided to the gate to form the depletion zone, P-channel depletion MOSFETs are in the "OFF" state.
- When a "+ve" (positive) voltage is provided to the gate, N-channel enhancement MOSFETs are in the "ON" state.
- When "-ve" (negative) voltage is delivered to the gate, P-channel enhancement MOSFETs are in the "ON" state.

The Field Effect Transistor Chart

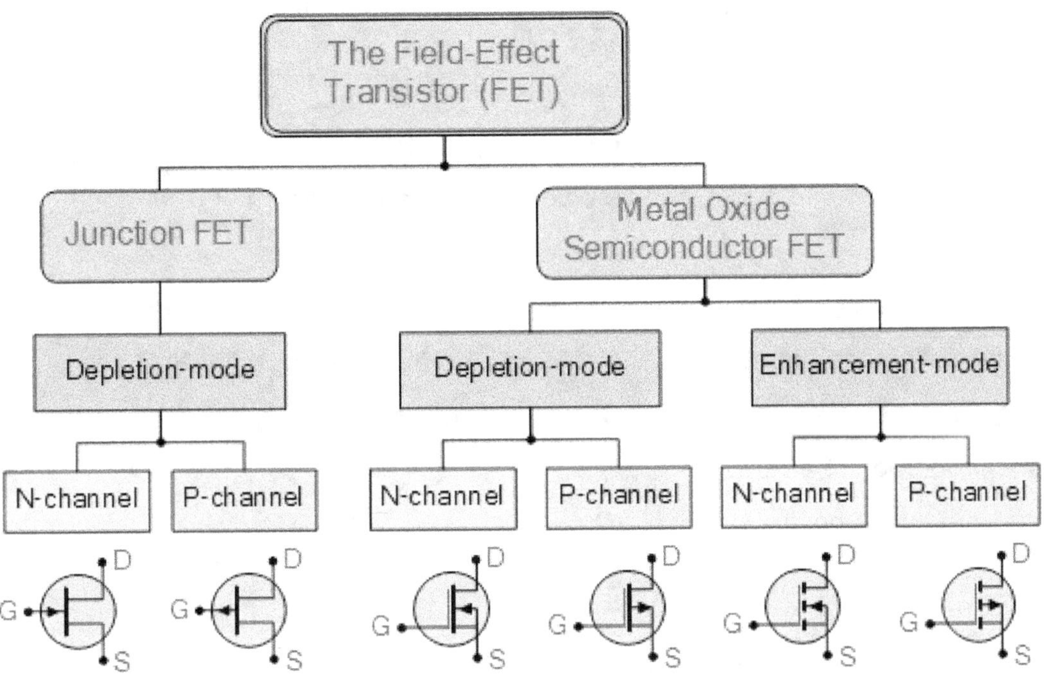

For both junction field effect transistor (JFET) and metal oxide semiconductor field effect transistor (MOSFET) types, the gate biasing is specified as follows:

Type	Junction FET		Metal Oxide Semiconductor FET			
	Depletion Mode		Depletion Mode		Enhancement Mode	
Bias	ON	OFF	ON	OFF	ON	OFF
N-channel	0V	-ve	0V	-ve	+ve	0V
P-channel	0V	+ve	0V	+ve	-ve	0V

Differences between a FET and a BJT

In electronic circuits, Field Effect Transistors can take the role of conventional Bipolar Junction Transistors. The table below provides a straightforward comparison of FETs and transistors, outlining both their benefits and drawbacks.

	Field Effect Transistor (FET)	Bipolar Junction Transistor (BJT)
1	Low voltage gain	High voltage gain

2	High current gain	Low current gain
3	Very high input impedance	Low input impedance
4	High output impedance	Low output impedance
5	Low noise generation	Medium noise generation
6	Fast switching time	Medium switching time
7	Easily damaged by static	Robust
8	Some require an input to turn it "OFF"	Requires zero input to turn it "OFF"
9	Voltage controlled device	Current controlled device
10	Exhibits the properties of a Resistor	
11	More expensive than bipolar	Cheap
12	Difficult to bias	Easy to bias

A list of complementary bipolar transistors that can be used as general-purpose low-current relay switches, LED and lamp drivers, amplifiers, and oscillators is provided below.

Complementary NPN and PNP Transistors

NPN	PNP	V_{CE}	$I_{C(max)}$	P_d
BC547	BC557	45v	100mA	600mW
BC447	BC448	80v	300mA	625mW
2N3904	2N3906	40v	200mA	625mW
2N2222	2N2907	30v	800mA	800mW
BC140	BC160	40v	1.0A	800mW
TIP29	TIP30	100v	1.0A	3W
BD137	BD138	60v	1.5A	1.25W
TIP3055	TIP2955	60v	15A	90W